Bandwidth Reduction of Stimulated Brillouin Scattering and Applications in Optical Communication

Bandwidth Reduction of Stimulated Brillouin Scattering and Applications in Optical Communication

Von der Fakultät für Elektrotechnik, Informationstechnik, Physik
der Technischen Universität Carolo-Wilhelmina zu Braunschweig

zur Erlangung des Grades eines Doktors
der Ingenieurwissenschaften (Dr.-Ing.)

genehmigte Dissertation

von Stefan Preußler

aus Riesa

eingereicht am: 12.05.2016

mündliche Prüfung am: 20.09.2016

1. Referent: Prof. Dr. rer. nat. Thomas Schneider
2. Referent: Prof. Avi Zadok

Druckjahr: 2016

Bibliografische Information der Deutschen Nationalbibliothek
Die Deutsche Nationalbibliothek verzeichnet diese Publikation in der
Deutschen Nationalbibliografie; detaillierte bibliografische Daten
sind im Internet über http://dnb.d-nb.de abrufbar.
1. Aufl. - Göttingen: Cuvillier, 2016
Zugl.: (TU)Braunschweig, Univ., Diss., 2016

Dissertation an der Technischen Universität Braunschweig,
Fakultät für Elektotechnik, Informationstechnik, Physik

© CUVILLIER VERLAG, Göttingen 2016
Nonnenstieg 8, 37075 Göttingen
Telefon: 0551-54724-0
Telefax: 0551-54724-21
www.cuvillier.de

ISBN 978-3-7369-9366-2
eISBN 978-3-7369-8366-3

Kurzfassung

Die stimulierte Brillouin Streuung ist auf Grund ihres sehr niedrigen Schwellwertes der dominanteste nichtlineare Effekt in optischen Einmodenfasern. Ihre einzigartigen spektralen Eigenschaften, insbesondere die geringe Bandbreite von 20–30 MHz, ermöglichen eine Vielzahl von Anwedungen, z.B. optische Spektralanalyse, Verzögerung und Speicherung von Licht, verteilte Sensoren sowie optische Signalverarbeitung. Eine weitere Verringerung der Brandbreite der stimulierten Brillouin Streuung führt zu einer signifikanten Verbesserung der Anwendungen und eröffnet neue Möglichkeiten.

In dieser Arbeit werden mehrere Verfahren zur Verringerung der Brillouin Bandbreite untersucht. Zu Beginn wird die Abhängigkeit der Brillouin Bandbreite von den Fasereigenschaften und der Pumpleistung, sowie der Verwendung einer Blende im Frequenzbereich erläutert. Weitere Untersuchungen umfassen die Überlagerung des Brillouin Gewinnes mit zwei Brillouin Verlusten, sowie die Verwendung eines mehrstufigen Brillouin Systems. Dabei ist es möglich die Brillouin Bandbreite auf bis zu 3 MHz zu reduzieren. Das entspricht ca. 15% der normalen Bandbreite.

Mit Hilfe der reduzierten Bandbreite kann die Auflösung eines Brillouin basierten optischen Spektrumanalysators signifikant erhöht werden. Zusätzlich wird zur Verbesserung des Dynamikbereiches die Polarisationsabhängigkeit der stimulierten Brillouin Streuung verwendet. Durch eine polarisationsgestützte Vorfilterung und anschließende heterodyne Überlagerung kann die Auflösung im Vergleich zu einem herkömmlichen Gitter basierten Spektrumanalysator um 6 Größenordnungen erhöht werden. Des Weiteren wird eine neue Methode für die Speicherung von optischen Datenpaketen, mit dem Namen Quasi-Licht-Speicherung, eingeführt. Das Verfahren beruht auf der Zeit-Frequenz-Kohärenz zwischen den Signalen. Die maximal erreichbare Speicherzeit ist dabei umgekehrt proportional zur Brillouin Bandbreite. Mit Hilfe der verschiedenen Methoden zur Bandbreitenreduzierung kann die maximale Speicherzeit auf bis zu 160 ns in einem einstufigen System und auf bis zu 500 ns in einem rückgekoppelten System erhöht werden.

Abschließend werden die schmalbandigen Filtereigenschaften der Brillouin Streuung für die Bearbeitung von optischen Frequenzkämmen verwendet. Das Ausschneiden einer bestimmten Anzahl von Linien aus dem Kamm führt dabei entweder zu einer schmalbandigen abstimmbaren Laserquelle, der Erzeugung von mm- und THz-Wellen mit extrem schmaler Linienbreite und sehr niedrigem Phasenrauschen oder durch die Weiterverarbeitung mit zwei gekoppelten Modulatoren zur Erzeugung von nahezu ideal geformten Nyquistpulsen.

Abstract

Stimulated Brillouin scattering is, because of its low threshold, the most dominant nonlinear effect in single mode optical fibers. Its unique spectral characteristics, especially the narrow bandwidth of 20–30 MHz, enable numerous applications, including optical spectrum analysis, delay and storage of light, distributed sensing and optical signal processing. Most of them would benefit from a reduction of the Brillouin gain bandwidth.

This work will introduce several methods for significant reduction of the Brillouin gain bandwidth. First, the dependence on the fiber properties and pump powers are investigated, followed by the utilization of a frequency domain aperture. Further explorations include the superposition of the Brillouin gain with two Brillouin losses in a three pump wave system, as well as, the utilization of a multi stage Brillouin amplifier. As will be shown, the Brillouin gain bandwidth can be reduced significantly down to 3 MHz, which equals 15% of the normal bandwidth.

The reduced Brillouin gain bandwidth considerably enhance the resolution of a Brillouin based optical spectrum analyzer. Additionally, the strong polarization dependence of stimulated Brillouin scattering is employed to enhance the dynamic range as well. Finally, by polarization assisted Brillouin prefiltering and subsequent heterodyne detection the resolution can be enhanced up to six orders of magnitude compared to a conventional grating based spectrum analyzer, utilizing just of the shelf equipment.

Further, a new technique for the storage of optical data packets, called Quasi-Light-Storage, is introduced. It is simply based on the time-frequency coherence between the signals. The maximum achievable storage time is inversely proportional to the Brillouin gain bandwidth. With several methods the maximum storage time could be enhanced up to 160 ns for a single stage system and up to 500 ns for a feedback system.

Finally, the narrow band filter characteristics of Brillouin scattering are employed for the processing of optical frequency combs. The extraction of a specific number of lines out of the comb leads to either a narrow band tunable laser source, the generation of mm- and THz-waves with ultra narrow linewidth and extremely low phase noise, or by further processing with two coupled modulators, to the generation of almost ideal shaped Nyquist pulses.

Acknowledgments

First of all, I wish to express my sincere gratitude to my supervisor Prof. Dr. rer. nat. Thomas Schneider for his encouragement and inspiration within the last years. He has taught me how to think about problems and find ways in solving them as a PhD student. I am grateful for his guidance in the basic theory, designing the experiments, analyzing the experimental data and improving my skills in writing scientific papers. Within numerous fruitful discussions almost every problem was solved and new ideas were developed. Without his support and assistance this work would not have been possible.

Moreover I would like to express my truly gratefulness to Prof. Avinoam Zadok from Bar-Ilan University in Israel. I admire his ability to quickly understand the essence of the concept and propose an original solution. He patiently taught me the theory of polarization pulling and guided me through the first experiments in this regard. It was an honor for me to work with him during his visit of our group. I enjoyed every time we met in the last years and discussing different topics with him.

Furthermore, I would like to thank all my former colleagues for their support, the stimulating discussions, the help in the laboratory and the very pleasant and relaxed working atmosphere. In particular I would like to thank Dr. Andrzej Wiatrek and Dr. Ronny Henker for the introduction into the laboratory and for the support in the challenging initial time, as well as Dr. Kambiz Jamshidi for fruitful discussions leading to further insights to the field. I would also like to thank Jens Klinger and Norman Wenzel for their active support in the laboratory. I am thankful for all my new colleagues and students i had in the last years, giving me the chance to improve my teaching skills and the ability to explain scientific facts.

Finally, and most of all, I would like to express my deep gratitude to my family and friends for their moral support and understanding over the last few years.

Contents

Acronyms

AW	AllWave fiber
AWG	arbitrary waveform generator
BOSA	Brillouin based optical spectrum analyzer
BPSK	binary phase shift keying
C	circulator
CW	continuous wave
DBG	dynamic Brillouin grating
DCF	dispersion compensating fiber
DFB	distributed feedback
DSF	dispersion shifted fiber
EDFA	erbium doped fiber amplifier
ESA	electrical spectrum analyzer
FBG	fiber Bragg grating
FL	fiber laser
FP	Fabry-Pérot
FSR	free spectral range
FWHM	full width at half maximum
FWM	four wave mixing
LD	laser diode
LDC	laser diode current controller
LIA	lock-in amplifier
LO	local oscillator
MLL	mode locked laser
MPF	microwave photonic filter
MZM	Mach-Zehnder modulator
OFDM	orthogonal frequency division multiplexing
OPM	optical power meter
OSA	optical spectrum analyzer
OSO	optical sampling oscilloscope
PBS	polarization beam splitter
PC	polarization controller

PD	photo diode
PDH	Pound-Drever-Hall
PM	phase modulator
PMD	polarization mode dispersion
PPA-SBS	polarization pulling assisted stimulated Brillouin scattering
PRBS	pseudo random bit sequence
QLS	Quasi-Light-Storage
SBS	stimulated Brillouin scattering
SNR	signal to noise ratio
SOP	state of polarization
SPM	self-phase modulation
SSMF	standard single mode fiber
SUT	signal under test
TEC	temperature controller
TIA	transimpedance amplifier
TLS	tunable laser source
TW	TrueWave fiber
VOA	variable optical attenuator
WDM	wavelength division multiplexing
WS	wave shaper
XPM	cross-phase modulation

1

Introduction

The field of optics and photonics has a long history, starting with the development of lenses by the ancient Egyptians and Mesopotamians, followed by theories on light and vision developed by ancient Greek philosophers. During the centuries perennial new phenomena were found and described, leading to the mathematical law of refraction, now known as Snell's law, in 1621. The investigation of the interaction of light and matter continued over centuries. In 1854, John Tyndall demonstrated to the Royal Society that light could be conducted through a curved stream of water, proving that a light signal could be bent. In 1880, Alexander Graham Bell invented his 'Photophone', which transmitted a voice signal on a beam of light. At the end of the 19th century the guidance of light through bent glass rods has been shown for the purpose of lighting in houses or as surgical lamp, as well as, for guiding light images in an attempt of early television. Although uncladded glass fibers were fabricated in the 1920s the field of fiber optics was not born until the 1950s when the use of a cladding layer led to considerable improvement in the fiber characteristics. These early fibers inhabit an extreme loss of more than $1000\,\mathrm{dB/km}$. Continuous engineering in purity of the glass and the fiber properties finally led to a loss of only $0.2\,\mathrm{dB/km}$ in the $1550\,\mathrm{nm}$ wavelength region. The availability of low-loss silica fibers in combination with the evolution of of laser devices, led not only to a revolution in the field of optical fiber communications but also to the advent of the new field of nonlinear fiber optics.

Light is an electromagnetic wave phenomenon described by the same theoretical principles that govern all forms of electromagnetic radiation. The Maxwell equations for a light wave propagating in a dielectric medium without free electronic charges or currents are well known [1, 2]:

$$\nabla \times \mathbf{E} = -\frac{\partial \mathbf{B}}{\partial t} \tag{1.1}$$

$$\nabla \times \mathbf{H} = \frac{\partial \mathbf{D}}{\partial t} \tag{1.2}$$

1

$$\nabla \cdot \mathbf{D} = 0 \tag{1.3}$$

$$\nabla \cdot \mathbf{B} = 0 \tag{1.4}$$

where $\mathbf{E}$ and $\mathbf{H}$ are the electric and magnetic field vectors and $\mathbf{D}$ and $\mathbf{B}$ are the corresponding electric and magnetic flux densities. $\nabla \times$ and $\nabla \cdot$ represent the curl and divergence vector operators. The relationship between the electric flux $\mathbf{D}$ and the electric field $\mathbf{E}$ depends on the electrical properties of the medium, described by the dielectric polarization $\mathbf{P}$. Within a dielectric medium the dielectric polarization is the macroscopic sum of of all electric dipole moments induced by the electric field. Correspondingly the relationship between magnetic flux density $\mathbf{B}$ and the magnetic field strength $\mathbf{H}$ depends on the magnetic properties of the material described by the magnetization $\mathbf{M}$, which is defined analogously to the dielectric polarization. The relations between the flux densities and the field strengths are given by [1, 2]:

$$\mathbf{D} = \varepsilon_0 \mathbf{E} + \mathbf{P} \tag{1.5}$$

$$\mathbf{B} = \mu_0 \left(\mathbf{H} + \mathbf{M} \right) \tag{1.6}$$

where ε_0 is the vacuum permittivity, μ_0 is the vacuum permeability. For a nonmagnetic medium, such as an optical fiber, $\mathbf{M} = 0$. In a linear, non-dispersive, homogeneous and isotropic medium the vectors $\mathbf{P}$ and $\mathbf{E}$ are parallel and proportional to each other, leading to [2]:

$$\mathbf{P} = \varepsilon_0 \chi \mathbf{E} \tag{1.7}$$

where the scalar constant χ is called dielectric susceptibility. The substitution of Eqn. 1.7 in Eqn. 1.5 shows that also $\mathbf{D}$ and $\mathbf{E}$ are parallel and proportional:

$$\mathbf{D} = \varepsilon \mathbf{E} \tag{1.8}$$

where the scalar variable

$$\varepsilon = \varepsilon_0 \left(1 + \chi \right) \tag{1.9}$$

is the permittivity of the medium. The relative permittivity $\varepsilon/\varepsilon_0 = 1 + \chi$ is also referred to as the dielectric constant of the medium. In order to find the wave equation for the propagation of light in a medium Maxwell's equations can be used. Taking the curl of Eqn. 1.1 and using Eqns. 1.2, 1.5 and 1.6, $\mathbf{B}$ and $\mathbf{D}$ can be eliminated in the favor of $\mathbf{E}$ and $\mathbf{P}$, leading to [1]:

$$\nabla \times \nabla \times \mathbf{E} = -\frac{1}{c^2}\frac{\partial^2 \mathbf{E}}{\partial t^2} - \mu_0 \frac{\partial^2 \mathbf{P}}{\partial t^2} \tag{1.10}$$

where c is the speed of light in vacuum and the relation $\mu_0 \varepsilon_0 = 1/c^2$ was used. Now it is convenient to represent this equation in the frequency domain [1]:

$$\nabla \times \nabla \times \tilde{\mathbf{E}} = \varepsilon(\omega)\frac{\omega^2}{c^2}\tilde{\mathbf{E}} \tag{1.11}$$

where $\tilde{\mathbf{E}} = \int_{-\infty}^{\infty} \mathbf{E} \exp(j\omega t)\, \mathrm{d}t$ is the Fourier transform of $\mathbf{E}$ and $\varepsilon(\omega) = 1 + \tilde{\chi}$ is the frequency-dependent relative dielectric constant. As $\tilde{\chi}(\omega)$ is in general complex, so is $\varepsilon(\omega)$. Its real and imaginary parts can be linked to the refractive index $n(\omega)$ and the absorption coefficient $\alpha(\omega)$ by using the definition [1]:

$$\varepsilon = (n + j\alpha c/2\omega)^2. \tag{1.12}$$

The refractive index and attenuation of the fiber are related to the susceptibility by the relations [1]:

$$n(\omega) = 1 + \frac{1}{2}\mathrm{Re}\left[\tilde{\chi}(\omega)\right] \tag{1.13}$$

$$\alpha(\omega) = \frac{\omega}{nc}\mathrm{Im}\left[\tilde{\chi}(\omega)\right] \tag{1.14}$$

where Re and Im stands for the real and imaginary part, respectively. Further simplifications to Eqn. 1.11 can be made due to the low optical losses in an optical in the wavelength region of interest. Therefore, the imaginary part of $\varepsilon(\omega)$ is small in comparison to the real part and $\varepsilon(\omega)$ can be replaced by $n^2(\omega)$. Additionally the vector identity $\nabla \times \nabla \times \mathbf{E} \equiv \nabla(\nabla \cdot \mathbf{E}) - \nabla^2 \mathbf{E} = -\nabla^2 \mathbf{E}$ can be utilized, leading to the following expression in form of the Helmholtz equation [1]:

$$\nabla^2 \tilde{\mathbf{E}} + n^2(\omega)\frac{\omega^2}{c^2}\tilde{\mathbf{E}} = 0. \tag{1.15}$$

The nonlinear effects in optical fibers occur either due to intensity dependence of the refractive index of the medium or due to scattering phenomena. Fiber nonlinearities are important in optical communications, both as useful attributes and as characteristics to be avoided. For intense electromagnetic fields, a homogeneous material behaves like a nonlinear medium. Therefore, the material polarization $\mathbf{P}$ is no longer proportional to the electric field $\mathbf{E}$ and can be decomposed by a Taylor series [3]:

$$\mathbf{P} = \varepsilon_0\chi^{(1)}\mathbf{E} + \varepsilon_0\chi^{(2)}\mathbf{E}^2 + \varepsilon_0\chi^{(3)}\mathbf{E}^3 + \ldots = \mathbf{P}^L + \mathbf{P}^{NL} \tag{1.16}$$

where ε_0 is the permittivity of vacuum and $\chi^{(k)}$ is the k-th order susceptibility. The dominant contribution to $\mathbf{P}$ is the linear susceptibility $\chi^{(1)}$, as described previously. The second order susceptibility $\chi^{(2)}$ is responsible for second-harmonic as well as sum-frequency generation. A medium, which lacks inversion symmetry at the molecular level, has non-zero second order susceptibility. However, for a symmetric molecule, like silica, $\chi^{(2)}$ vanishes. Therefore, optical fibers do not exhibit second order nonlinear refractive effects. Obviously the third order susceptibility $\chi^{(3)}$ is responsible for lowest order nonlinear effects in fibers [1]. The refractive index in the presence of this type of nonlinearity can be represented as [4]:

$$n = n_0 + n_2 I \tag{1.17}$$

with $I = \frac{1}{2}c\varepsilon_0 n_0 E_0^2$ and n_0 as the linear or low-intensity refractive index. The optical constant n_2, that characterizes the strength of the optical nonlinearity, can be expressed by [4]:

$$n_2 = \frac{3}{2}\frac{\chi^{(3)}}{c\varepsilon_0 n_0} \tag{1.18}$$

The power dependence of the refractive index is responsible for the Kerr-effect. Depending on the type of input signal, the Kerr-nonlinearity manifests itself in different effects, including:

- self-phase modulation (SPM): implies that the nonlinear phase modulation is self-induced. In principle, the higher intensity portions of an optical pulse encounter a higher refractive index of the fiber compared with the lower intensity portions while it travels through the fiber. Since the intensity of the signal is time varying it generates a time varying refractive index in the medium. The leading edge will experience a positive and trailing edge a negative refractive index gradient. This temporally varying index change results in a temporally varying phase change, leading to different phase shift of the different parts of the pulse. The final result is frequency chirping, where the rising edge of the pulse is frequency shifted in the upper side, whereas the trailing edge experiences a shift in lower side. Hence, the primary effect of SPM is to broaden the spectrum of the pulse and simultaneously keeping the temporal shape unaltered. With an sufficiently high amount of dispersion the pulse will be temporal broadened, but with equally distributed portions of SPM and dispersion, a soliton is formed which is neither dispersed in the temporal domain nor in the frequency domain. The SPM effects are more pronounced in systems with high-transmitted power because the chirping effect is proportional to transmitted signal power.

- cross-phase modulation (XPM): is a similar effect to SPM, but involves two or more optical pulses, for instance at different wavelengths, that can interact with each other via the alteration of the intensity dependent refractive index. XPM is always accompanied by SPM and occurs because the nonlinear refractive index seen by an optical wave depends not only on the intensity of that wave but also on the intensity of the other co-propagating waves. Therefore, XPM converts power fluctuations in a particular wavelength channel to phase fluctuations in other co-propagating channels. This effect can lead to asymmetric spectral broadening and distortion of the pulse shape, manifesting in degradation of system performance in optical communications.

- four wave mixing (FWM): The interaction of two or more light waves can lead to a second kind of $\chi^{(3)}$ nonlinearities, which involve an energy transfer between waves and not simply a modulation of the refractive index. If three optical fields with carrier frequencies $\omega_1, \omega_2, \omega_3$ co-propagate inside the fiber simultaneously, $\chi^{(3)}$ generates a fourth field with frequency ω_4, which is related to other frequencies by a relation, $\omega_4 = \omega_1 \pm \omega_2 \pm \omega_3$. Clearly there can be a large variety of such process, depending on the particular product of the four fields. SPM and XPM are significant mainly for high bit rate systems, but the FWM effect is independent of the bit rate and is critically dependent on the channel spacing and fiber dispersion. Decreasing the channel spacing increases the four-wave mixing effect and so does decreasing the dispersion.

In the global communications systems largely conventional silica fibers (type Corning SMF-28) used for data transmission. Since glass is a solidified liquid, there are small variations in material density and material composition. The size of the variations is in the range of the wavelength or less, which leads to minor changes in the refractive index. Such fluctuations in density, like impurities or thermal movement of particles in the material, are responsible for various kinds of scattering of the light in all directions of the space in an inhomogeneous material. There are basically two kinds of scattering:

- Elastic scattering, where the frequency of the scattered wave is identical to the incident wave,

- Inelastic scattering, where the scattered wave undergoes a frequency change with respect to the incident light wave.

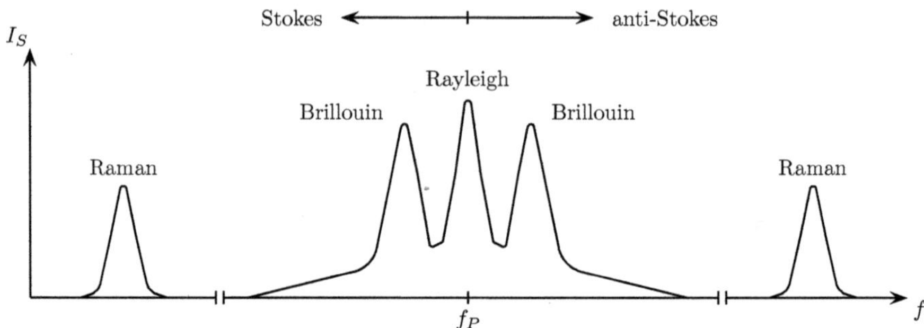

Figure 1.1.: Schematic representation of a typical spectrum of the scattering of light [4].

Under the most general circumstances, an incident light wave at the frequency f_P that is coupled in a scattering medium generates a typical spectrum as shown in Fig. 1.1. By definition, those components of the scattered light which are shifted to lower frequencies are called Stokes components. Additionally, energy is transferred from the pump wave to the stokes wave, which means energy loss for the pump wave. Contrary, components that are shifted to higher frequencies are referred to as anti-Stokes components. Here the pump wave is accompanied by an energy gain. In general, the spectrum of the scattered light consists of the following components [4, 5]:

- Rayleigh scattering, caused by structural variations in the density of the medium. It is known as quasi-elastic scattering, because no frequency shift is caused. The scattering from the sides of the Rayleigh scattering is the result of fluctuations in positioning of the anisotropic molecules. Due to the very rapid molecular reorientation these components are spectral very broad.

- Raman scattering, which is formed by the interaction of light with the vibrational states of the molecules of the medium. Equivalently, Raman scattering can be described as the scattering of light by optical phonons. Thereby, the scattered light is spectral shifted by several tens of THz and a spectral width with a magnitude in the THz range.

- Brillouin scattering, the scattering of light by sound waves, which have their origin in density variations of the material. Brillouin scattering can also be considered as the scattering of light by acoustic phonons. Thereby, the frequency shift at telecommunications wavelengths is in the range of 10–15 GHz and spectral width is in the range of 10–30 MHz.

In principle, all scattering processes occur at arbitrary frequencies f_P of the incident light wave, and are therefore not bound to narrow frequency ranges. All mentioned scattering processes can occur as [5]:

- Spontaneous scattering, which dominates at low power densities of the pump light wavelength. The scattered light is non-directional and the intensity is proportional to the pump wave. All spontaneous scattering processes have in common that the intensity of the scattering is proportional to the frequency f_P^4 or the wavelength λ_P^{-4} of the pump wave.

- Stimulated scattering, where the power of a wave is amplified at the scattering frequency by the power of the pump wave. The emission of the scattered light occur, depending on the direction of propagation of the pump wave, in a preferred direction. The intensity of the scattered light is strongly non-linearly dependent from that of the pump light. The intensities of the stimulated scattering processes are, in contrast to the spontaneous scattering processes, only proportional to f_P or λ_P^{-1} of the pump wave.

In the red and near infrared spectral range of 600 nm to 1550 nm, the Rayleigh scattering forms the theoretical limit of attenuation. After the previous classification Rayleigh scattering is a spontaneous scattering process. If additional refractive index fluctuations are induced by spatially variable thermal effects due to light absorption in the medium, then the Rayleigh scattering will be increased. This process may be referred to as stimulated Rayleigh scattering and can play a role in high-power fiber lasers [6, 7].

Raman scattering is exploited in many applications [8]. One classical application is the fiber based Raman laser [9], which can be also realized on silicon waveguides [10, 11]. Additional, Raman scattering is utilized as amplifier in optical fibers. The Raman amplification may occur at any wavelength as long as an appropriate pump laser is available. Raman amplification may be realized as a continuous amplification along the fiber which let the signal never become too low. A Raman amplifier is bidirectional in nature and more stable. It has been shown, that Raman amplifiers have several advantages compared to Erbium doped fiber amplifiers, particularly when the transmission fiber itself is used as a Raman amplifier [12]. Especially the advantages of self phase matching and broad gain-bandwidth lead to advantages in wavelength division multiplexed systems [13]. However, a more detailed description about Raman scattering including the physical process, gain, spectral shape and threshold can be found in [4, 5, 14].

The effect of Brillouin scattering, including applications, will be explained more detailed in the following chapter. Therefore, chapter 2 starts with the physical process of spontaneous and stimulated Brillouin scattering, followed by the main parameters. Further a detailed analysis of the polarization dependence is given. The remainder of the chapter focuses on possible applications of Brillouin scattering. As will be shown, most of the applications would benefit from a narrowed Brillouin gain bandwidth.

Accordingly, several methods for the Brillouin gain bandwidth reduction are introduced in chapter 3, accompanied with experimental evidence. First, an investigation of environmental influences and fiber properties on the bandwidth will be given. Afterward the bandwidth reduction by a multi stage system, the superposition of a Brillouin gain with two losses and the utilization of a frequency domain aperture will be introduced and explained in detail. The rest of the thesis will continue with selected applications in optical communication.

Chapter 4 will first give an introduction to classical optical spectrum analysis, followed by the utilization of polarization pulling assisted Brillouin scattering accompanied by a significant increase of the resolution through gain bandwidth narrowing. Furthermore, the Brillouin based optical spectrum analyzer will be connected with the technique of heterodyne detection to achieve unprecedented resolution.

Chapter 5 will introduce the delay and storage of light in general and subsequently focus on a new method called Quasi-Light-Storage, where the nonlinear effect of stimulated Brillouin scattering is exploited for the multiplication of a frequency comb with the power spectral density of a signal, leading to multiple time delayed copies of the signal. Through the utilization of the bandwidth narrowing methods the maximum storage time can be enhanced significantly. The basic theory, as well as possibilities and limits will be given.

The 6-th chapter utilizes Brillouin scattering as an narrow band optical filter for the processing of an optical frequency comb. Depending on the number of extracted lines, two main applications will be described. First the generation of high quality mm- and THz-waves including wireless transmission, and second, the generation of almost ideal sinc-shaped Nyquist pulse sequences. The last chapter will summarize the thesis and give an outlook for further enhancements and future applications.

2

Brillouin Scattering

Brillouin light scattering is generally referred to as inelastic scattering of an incident optical wave field by thermally excited elastic waves in a sample. The theoretical prediction of spontaneous light scattering on thermally excited acoustic waves was carried out in 1922 by namesake Léon Brillouin [15]. Leonid Mandelstam is believed to have recognized the possibility of such scattering as early as 1918, but he published his idea only in 1926 [16]. The experimental confirmation of spontaneous Brillouin scattering in liquids and crystals was performed in 1930 by Gross using a lamp as the light source [17]. After invention of the laser the effect of stimulated Brillouin scattering was first observed in 1964 by Chiao et al. [18]. In addition, the development of low-loss glass fibers was pushed by Corning and the Bell Laboratories in the 1970s [19], resulting in increased research towards optical-fiber transmission systems [20]. However, Brillouin scattering in optical fiber and especially the threshold of the stimulated process is well known since the early 1970s [21, 22]. The significance of stimulated Brillouin scattering (SBS) in fibers for optical communication systems was first described in detail by D. Cotter in 1983 [23]. Since then, the development of fibers and systems stride on and new insights were obtained [24].

Since its discovery, the SBS has affected optical systems in different ways. On the one hand it drastically limits the maximum transmittable power in standard single mode fibers [25]. Therefore, over the past decades several approaches for avoiding the detrimental effects of SBS or increasing the threshold were investigated, including the artificial broadening of the pump laser spectrum [26, 27], theoretically the utilizing of a single narrow band fiber Bragg grating along the fiber [28], isolators in the transmission system [29], changing the effective mode area and the doping of the fiber [30] as well as using fibers with different Brillouin shifts in the systems [31].

But on the other hand, stimulated Brillouin scattering offers a variety of applications in different fields. Besides the high resolution spectrum analysis, optical delay and quasi-light storage as well as general signal generation and processing, as will be discussed in the forthcoming chapters, stimulated Brillouin scattering in fibers can be utilized for thermal and strain sensing [32–34] and ultra-narrow linewidth light sources [35, 36]. Applications from other fields of science include the analysis of bulk viscosity of liquids [37], the measurement of temperature of water and even the ocean [38], microscopic imaging [39] and even characterization of elastic properties of materials in geoscience [40].

This chapter will give a detailed overview about the fundamentals and the physical process of spontaneous and stimulated Brillouin scattering. Subsequently, the coupled energy equations and key parameters including gain, bandwidth, threshold and shift are discussed. Additionally, the chapter will show the polarization attributes of stimulated Brillouin scattering and give an introduction to the polarization pulling effect. Finally, different applications of stimulated Brillouin scattering are discussed.

2.1. Spontaneous Brillouin Scattering

The scattering of light is caused by inhomogeneities of refractive index in the media. In real inhomogeneous propagation medium static fluctuations of the local dielectric permittivity lead to elastic scattering of light in all directions. In optical fibers, thermally reasoned lattice vibrations (phonons) produce already density fluctuations in the medium that spread with velocity of sound. The spontaneous Brillouin scattering refers to the scattering of a light wave with the frequency f_P on these thermally excited acoustic waves in a medium, as can be seen in Fig. 2.1 [5].

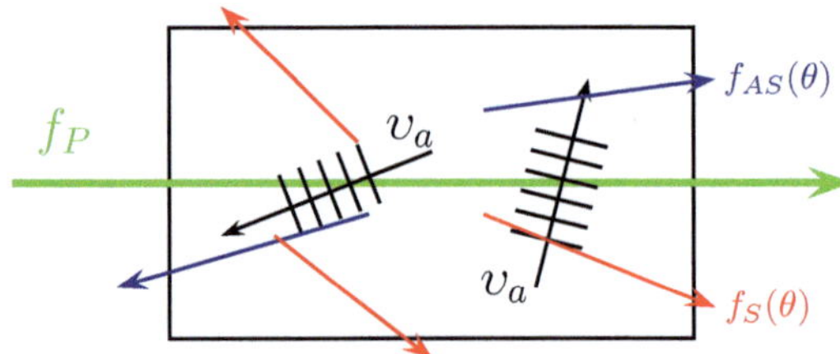

Figure 2.1.: Scattering process of spontaneous Brillouin scattering.

The sound waves propagate in the medium with the velocity v_a and produce a periodic modulation of the refractive index of a medium due to the photoelastic effect. The diffraction and reflection of light waves on stationary periodic refractive index gratings is known in wave theory as Bragg reflection [2].

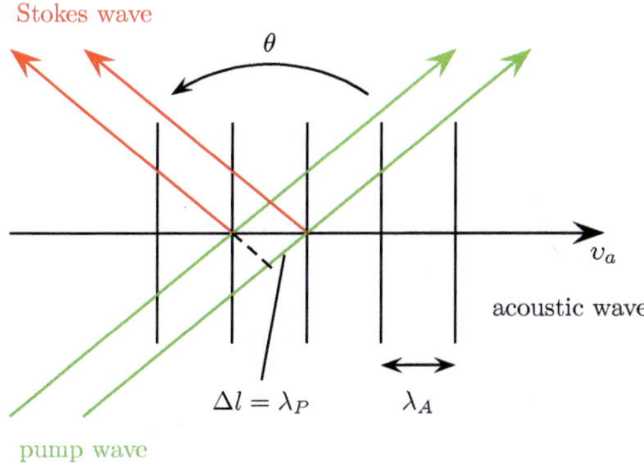

Figure 2.2.: Reflection of a light wave on a sound wave, which is represented by their phase fronts with the same density or refractive index gradient. Partial reflections of the pump light wave at the same refractive index gradient with the distance λ_A of the sound wave will superimpose constructively to a scattered Stokes wave when the entire path difference is $\Delta l = \lambda_P$.

The wavelength of the sound wave λ_A determines the spatial period of the refractive index modulation. At such a periodic refractive index a pump wave with the vacuum-wavelength λ_P is reflected when the Bragg condition is satisfied:

$$\lambda_P = 2n_P \lambda_A \sin \frac{\theta}{2} \tag{2.1}$$

where n_P is the optical refractive index of the medium of the incident pump wave, and θ the angle between the propagation direction of the incident and the reflected beam, as can be seen in Fig. 2.2. The propagation direction of the sound wave in isotropic media is always perpendicular to the bisecting line of θ. The scattered light wave experiences through reflection on the acoustic wave, which propagates with the speed of sound v_a, a frequency shift due to the Doppler effect:

$$f_S = f_P \left(1 - 2n \frac{v_a}{c_0} \sin \frac{\theta}{2} \right) = f_P - f_B(\theta) \tag{2.2}$$

$$f_{AS} = f_P \left(1 + 2n \frac{v_a}{c_0} \sin \frac{\theta}{2} \right) = f_P + f_B(\theta) \tag{2.3}$$

wherein at the same scattering angle θ a Stokes wave of smaller frequency f_S is generated

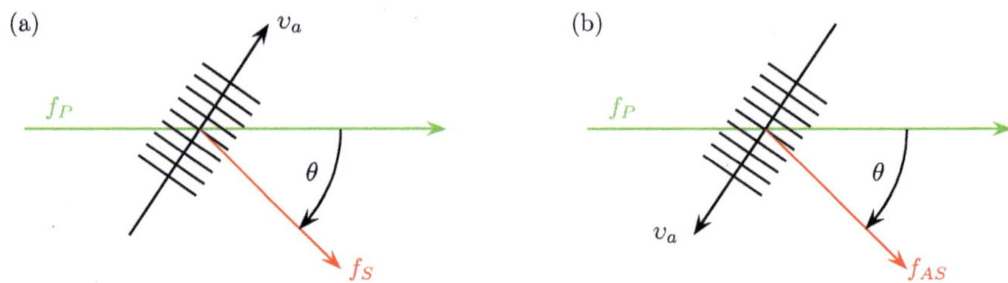

Figure 2.3.: Depending on the propagation direction of the sound wave v_a with respect to the pump light wave with the frequency f_P, Brillouin scattering generates at a scattering angle θ a Stokes wave (a) with a smaller frequency f_S or an anti-Stokes wave (b) with a higher frequency f_{AS} with respect to f_P.

when the acoustic wave moves away from the pump wave, and an anti-Stokes wave with greater frequency f_{AS} is generated when the sound wave moves towards the pump light wave, as can be seen in Fig. 2.3 [5].

The maximum frequency offset occurs during scattering in the reverse direction with $\theta = \pi$, and is referred to as material-specific Brillouin frequency shift f_B or shortly Brillouin frequency with the angular frequency Ω_B:

$$f_B = 2n\frac{v_a}{c_0}f_P, \qquad \Omega_B = 2\pi f_B \tag{2.4}$$

A comparison with Eqn. 2.1 shows that for $\theta = \pi$ the Brillouin frequency shift f_B is equal to the acoustic frequency f_A of the sound wave. The Brillouin shift in fused silica with a refractive index of $n = 1.445$ and a velocity of sound of $v_a = 5960\,\mathrm{m/s}$ at a pump wavelength of $\lambda_P = 1550\,\mathrm{nm}$, or $f_P = 193.4\,\mathrm{THz}$, is calculated to be $f_B = 11.1\,\mathrm{GHz}$. Which is caused by an acoustic wave with the frequency $f_A = f_B$. The Brillouin frequency shift f_B is typically three orders of magnitude smaller than the frequency offset of Raman scattering. While the Raman frequency shift is independent of the pump frequency and a material constant, the Brillouin frequency shift is directly proportional to the pump frequency f_P.

Spontaneous Brillouin scattering on thermally excited sound waves also takes place in optical fibers. Due to the waveguide properties and the propagation of light in the z-direction, only the forward scattering in the same direction and back scattering in the opposite direction with respect to the pump wave that is guided in the core are relevant. A lateral scattering with large angle with respect to the fiber axis is not guided in the core, but radiated into the jacket and into the environment. The most important process is the back scattering (scattering angle $\theta = \pi$) of a pump light wave on a longitudinal sound wave propagating in

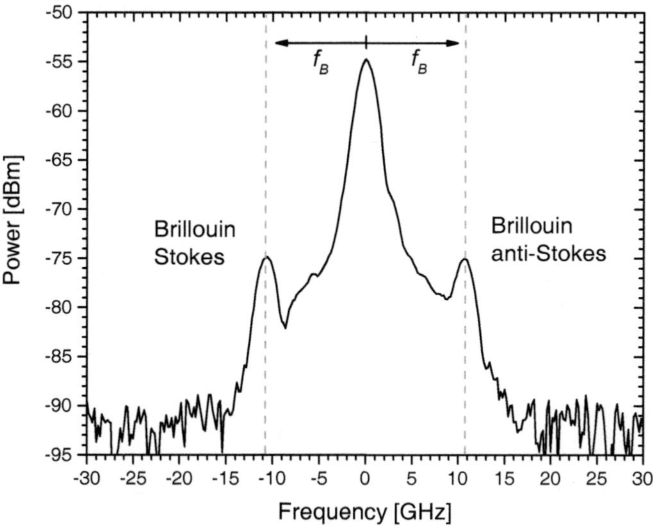

Figure 2.4.: Measured spectral distribution of spontaneously scattered Brillouin Stokes and anti-Stokes wave in a 50 km long standard single mode fiber. The center line shows the Rayleigh back scattering of the pump wave.

the z-direction. For a pump wave with the frequency f_P with propagation in +z-direction, a sound wave that travels in the same direction causes a back scattering the pump into a Stokes wave with smaller frequency $f_S = f_P - f_A$. An oppositely propagating acoustic wave in -z-direction is responsible for the back scattering in the anti-Stokes wave with higher frequency $f_{AS} = f_P + f_A$. The acoustic frequency f_A must satisfy the Bragg condition. Such scattered light is guided in the core and can accumulate over long lengths. This scattering process is usually referred to as spontaneous Brillouin scattering in optical fibers. Exemplary, the measurement of spontaneous Brillouin scattering in a 50 km long standard single mode fiber (SSMF) can be found in Fig. 2.4, clarifying the Stokes- and anti-Stokes waves with respect to the Rayleigh back scattered pump wave.

Furthermore, caused by light scattering in the forward direction ($\theta = 0$) by acoustic waves that moves transverse to the fiber axis, a phase and polarization modulation of the pump wave will be generated. This process is referred to as guided acoustic wave Brillouin scattering, to distinguish it from the ordinary spontaneous Brillouin scattering. However, it will find no further consideration in this work. In principle, just a small part of the optical power is scattered by the spontaneous Brillouin scattering, so that this effect per se is of little technical importance. But, the Stokes wave is amplified during further propagation along the fiber through stimulated Brillouin scattering. This allows the power of the Stokes wave to reach the same order as the pump wave.

2.2. Stimulated Brillouin Scattering

As described previously, a pump wave with the frequency f_P will generate spontaneous Brillouin scattering in a fiber through thermally excited acoustic waves. A small part of this is exactly counter propagating to the pump wave with a frequency of f_S and will be amplified due to the Brillouin gain. This process, illustrated in Fig. 2.5(a), arises basically from noise and can be called stimulated-amplified spontaneous Brillouin scattering and can be seen as a SBS-Generator [4]. This effect mainly limits the maximum transmittable power for an optical fiber and will be discussed in detail in section 2.5.

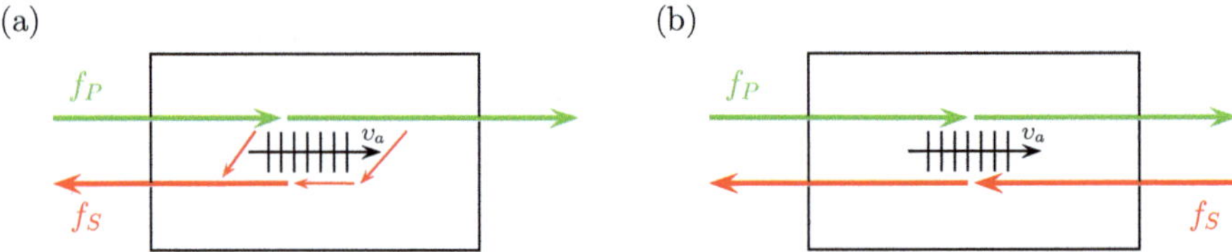

Figure 2.5.: Scattering process for stimulated Brillouin scattering initiated by spontaneous Brillouin scattering (a) or a counter propagating wave (b).

On the other hand, stimulated Brillouin scattering can be initiated by a counter propagating wave injected into the fiber at the frequency $f_S = f_P - f_B$, as shown in Fig. 2.5(b). This process is called Brillouin amplifier. The interference between pump and Stokes wave leads to a beating which envelope frequency is identical to the difference frequency between the two waves and moves in the direction of the higher frequency, as can be seen in Fig. 2.6. This intensity modulation will be translated into a density modulation of the medium via electrostriction. Electrostriction appears in materials such as silica, which tends to squeeze when a strong electric field is present. This phenomenon is associated to the tendency of molecules to move or reorientate in presence of an electric field in order to minimize the potential energy. This mechanical deformation as a result of the electric field causes a periodical density variation or an acoustic wave in the medium. With the density modulation of the material, its refractive index is modulated as well. This leads to a periodic grating in the material which fulfills Eqn. 2.4. With increasing pump power the intensity of the interference wave also increases, which in turn leads to an increase of the grating and thus

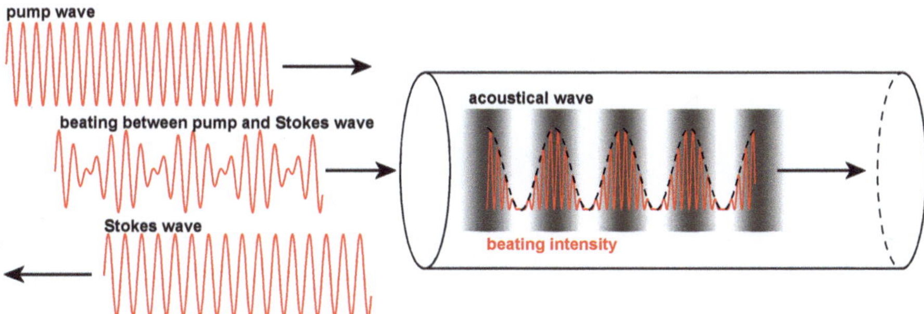

Figure 2.6.: Formation of the acoustic wave in the medium from the superposition of pump and Stokes wave for a fixed time (t = const).

to an exponential increase of the power of the backscattered wave. The signal wave, which is counter propagating to the pump wave, accumulates the scattered pump energy and is thus amplified. In general, the Brillouin scattering is referred to as stimulated when a change in power of the pump wave cause a change in power of the signal or Stokes wave. If an electromagnetic wave at the frequency of the anti-Stokes wave is coupled counter propagating to the pump wave into the fiber, the direction of energy flow from the higher to the lower frequency is identical to the above described process. Therefore, signals in the anti-Stokes region are attenuated depending on the pump power.

For an electrostrictive material the nonlinear polarization, as a function of the change occurred in the susceptibility, can be expressed by [4]:

$$\mathbf{P}_i^{NL} = \Delta\varepsilon \mathbf{E}_i\left(\mathbf{r}, t\right) = \frac{\gamma_e}{\rho_0}\Delta\rho \mathbf{E}_i\left(\mathbf{r}, t\right) \tag{2.5}$$

where γ_e is the electrostrictive coefficient, ρ_0 the average material density, and ε the dielectric permittivity and $\Delta\rho$ the function of the density wave in the medium. The process of SBS is classically described by the interaction of a pump wave $\mathbf{E}_P$, a Stokes signal wave $\mathbf{E}_S$ and an acoustic wave $\Delta\rho$. Since wave propagation in optical fibers only occurs along the axial direction, the SBS process can be appropriately modeled as a one-dimensional interaction between two counter propagating light waves and an acoustic wave that is co-propagating with respect to the pump wave [4]:

$$\mathbf{E}_P = \mathbf{e}_P \frac{1}{2} E_P\left(z,t\right) e^{j(\omega_P t - k_P z)} + c.c.$$

$$\mathbf{E}_S = \mathbf{e}_S \frac{1}{2} E_S\left(z,t\right) e^{j(\omega_S t + k_S z)} + c.c. \tag{2.6}$$

$$\Delta\rho = \frac{1}{2} A\left(z,t\right) e^{j(\omega_B t - k_B z)} + c.c.$$

where $\mathbf{e}_i = \mathbf{E}_i / |\mathbf{E}_i|$ are the normalized polarization vectors for the optical fields. A more detailed analysis on the polarization properties of SBS will be given section 2.4. The temporal and spatial evolution of these waves is subjected to their optical or acoustic perturbed wave equations. In an electrostrictive, isotropic and homogeneous medium, the three interacting waves using the following equations [5]:

$$\nabla^2 \mathbf{E}_P - \frac{n^2}{c^2}\frac{\partial^2 \mathbf{E}_P}{\partial t^2} = \mu_0 \frac{\partial^2 \mathbf{P}_P^{NL}}{\partial t^2}$$

$$\nabla^2 \mathbf{E}_S - \frac{n^2}{c^2}\frac{\partial^2 \mathbf{E}_S}{\partial t^2} = \mu_0 \frac{\partial^2 \mathbf{P}_S^{NL}}{\partial t^2} \tag{2.7}$$

$$v_a^2 \nabla^2 \left(\Delta\rho\right) - \frac{\partial^2 \Delta\rho}{\partial t^2} - \Gamma \frac{\partial \Delta\rho}{\partial t} = \nabla \cdot \mathbf{F}$$

where Γ is the acoustic damping coefficient and P_i^{NL} are the nonlinear polarization fields, given by Eqn. 2.5. For the given equations $\Delta\rho$ denotes the density variations rather than the pressure variations to describe the acoustic wave. These equations can be all coupled through the electrostrictive force F using the material constitutive relations given by [4]:

$$\nabla \cdot \mathbf{F} = \nabla p_{el} = -\frac{\gamma_e}{2}\nabla^2 \left\langle E^2 \right\rangle \tag{2.8}$$

where p_{el} is the electrostrictive pressure. Consequently, the nonlinear polarization terms for the pump and Stokes wave can be expressed by:

$$\mathbf{P}_P^{NL} = \mathbf{e}_P \frac{1}{2}\frac{\gamma_e}{\rho_0} A\left(z,t\right) E_S\left(z,t\right) e^{j(\omega_P t - k_P z)} + c.c.$$

$$\tag{2.9}$$

$$\mathbf{P}_S^{NL} = \mathbf{e}_S \frac{1}{2}\frac{\gamma_e}{\rho_0} A^*\left(z,t\right) E_P\left(z,t\right) e^{j(\omega_S t + k_S z)} + c.c.$$

In the case of $\mathbf{P}^{NL} = 0$, the two light waves $\mathbf{E}_P$ and $\mathbf{E}_S$ from Eqn. 2.7 are not coupled and their intensities remain constant during propagation. Therefore, the solution is given simply by plane waves. In the case of $\mathbf{P}^{NL} \neq 0$, the nonlinear polarization acts as a coupling term for the fields $\mathbf{E}_P$ and $\mathbf{E}_S$. In most cases $\left|\mathbf{P}^{NL}\right| \ll \left|\mathbf{P}^{L}\right|$, so that the effect of nonlinear polarization is modest for propagation distances of a few optical wavelengths. Therefore, the linear solutions for the wave fields can be utilized, applying the slowly varying envelope approximation [41]. With this assumption the system of differential coupled equations is reduced to a set of scalar equations [5]:

$$\frac{\partial}{\partial z}E_P - \frac{n}{c_0}\frac{\partial}{\partial t}E_P + \frac{\alpha}{2}E_P = -j\frac{k_S\gamma_e}{4\varepsilon\rho_0}AE_S$$

$$\frac{\partial}{\partial z}E_S + \frac{n}{c_0}\frac{\partial}{\partial t}E_S - \frac{\alpha}{2}E_S = -j\frac{k_P\gamma_e}{4\varepsilon\rho_0}A^*E_P \qquad (2.10)$$

$$\frac{\partial}{\partial z}A + \frac{\omega_B - j\Gamma k_B^2}{\omega_B v_a}\frac{\partial}{\partial t}A + \frac{\Gamma k_B^2}{v_a}A = -\left(\mathbf{e}_P \cdot \mathbf{e}_S\right)j\frac{k_B\gamma_e}{4v_a^2}E_P E_S^*$$

Additionally, the optical loss of the medium has been introduced by the coefficient α. During the whole scattering process, both the energy and momentum must be conserved, thus the frequencies and the wave vectors of the pump, Stokes and acoustic wave are related by:

$$\omega_B = \omega_P - \omega_S$$
$$\mathbf{k}_B = \mathbf{k}_P - \mathbf{k}_S \qquad (2.11)$$

where ω_B is the angular frequency and k_B the wave vector of the acoustic wave. In general the vector characteristics of k_B can be neglected due to preferred longitudinal propagation direction of the acoustic wave in the fiber. With this system of equations, both static and dynamic SBS processes can be described. A solution for variable pulse shapes of the pump and Stokes wave is only possible numerically. For continuous wave signals, where $\partial/\partial t = 0$, the time derivative term in the third equation can vanish as assuming the steady-state conditions. Furthermore, optical waves can propagate over long distances within a fiber due to the low optical losses. Contrary, the hypersonic acoustic waves are attenuated rapidly. This leads to a usually very small mean free path of the phonons ($\sim 10^{-6}\,\mathrm{m}$). Thus the amplitudes of acoustic wave can be given by:

$$A = j\frac{k_B\gamma_e}{2\Gamma_B v_a}(\mathbf{e}_P \cdot \mathbf{e}_S)E_P E_S^* \qquad (2.12)$$

where $\Gamma_B = \Gamma k_B^2$ denotes the bandwidth of Brillouin scattering, depending directly on the acoustic damping coefficient which represents the average lifetime of the acoustic phonons inside the medium [1]. Finally, inserting the amplitude of the acoustic wave in Eqn. 2.10 leads to two coupled equations relating to the pump and Stokes waves:

$$\frac{\partial E_P}{\partial z} = -(\mathbf{e}_P \cdot \mathbf{e}_S)\frac{k_B k_P \gamma^2}{8\varepsilon\rho_0\Gamma_B v_a}\frac{E_P \left|E_S\right|^2}{1 - j\left(2\Delta\nu/\Delta\nu_B\right)} - \frac{\alpha}{2}E_P$$

$$\frac{\partial E_S}{\partial z} = -(\mathbf{e}_P \cdot \mathbf{e}_S)\frac{k_B k_S \gamma^2}{8\varepsilon\rho_0\Gamma_B v_a}\frac{E_S \left|E_P\right|^2}{1 - j\left(2\Delta\nu/\Delta\nu_B\right)} + \frac{\alpha}{2}E_S$$

$$(2.13)$$

where ν is the frequency offset between the pump and Stokes wave and ν_B is the Brillouin frequency shift given by Eqn. 2.4 and $\Delta\nu$ is a detuning frequency given by $(\nu - \nu_B)$. As can be seen from these equations, two main phenomenon will arise. The real part of the equation leads to an energy transfer and therefore to an optical gain (>0) or an optical loss (<0), and the imaginary part leads to a nonlinear phase shift. The equations can be further simplified taking the steady state conditions into account, where the life time of the acoustic wave is so short that it can be neglected when compared to the that of the pump wave. Additionally, the optical intensities $I_i = \frac{1}{2}n\varepsilon_0 c \left|E_i\right|^2$ are introduced for the pump and Stokes wave. The coupled intensity equations can be written as [1]:

$$\frac{\partial I_P}{\partial z} = -g_B\left(\Delta\nu\right) I_P I_S - \alpha I_P$$

$$(2.14)$$

$$\frac{\partial I_S}{\partial z} = -g_B\left(\Delta\nu\right) I_P I_S + \alpha I_S$$

where a new parameter g_B, the Brillouin gain coefficient, is introduced. The loss resonance generated by SBS for a counter propagating wave with the frequency ν_{AS} is mathematically described in the same manner as the Stokes scattering. The only difference is that the sign of the Brillouin gain coefficient is negative and the frequencies of the two waves are exchanged ($\nu_{AS} > \nu_P$). The spectral representation of the Brillouin loss is identical to the gain as depicted in Fig.2.7.

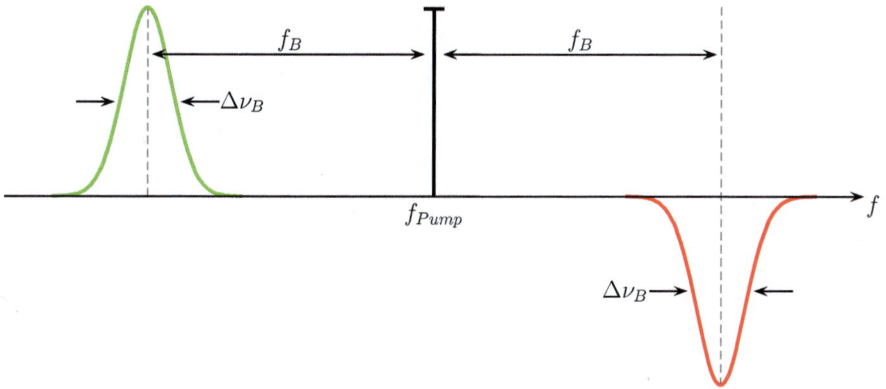

Figure 2.7.: The Brillouin gain and loss resonance with respect to the pump wave.

2.3. The Brillouin Gain

The spectral representation of the Brillouin gain coefficient is defined by [1, 4]:

$$g_B\left(\Delta\nu\right) = g_p \frac{\left(\Delta\nu_B/2\right)^2}{\left(\Delta\nu\right)^2 + \left(\Delta\nu_B/2\right)^2} \tag{2.15}$$

with $\Delta\nu_B$ being the Brillouin linewidth and g_P is the peak value of the Brillouin gain coefficient defined by [4]:

$$g_P = \frac{\eta_P 2\pi\gamma^2}{c\varepsilon_0^2\lambda_P^2\rho_0\Delta\nu_B v_a n} = \frac{\eta_P 2\pi n^7 p_{12}^2}{c\lambda_P^2\rho\Delta\nu_B v_a} \tag{2.16}$$

where λ_P denotes the pump wavelength and p_{12} is the longitudinal elasto-optic coefficient. The factor η_P is the polarization mixing efficiency given by $\eta_P = |\mathbf{e}_P \cdot \mathbf{e}_S|^2$. In bulk silica the measurable of g_P is approximately $5 \cdot 10^{-11}\,\mathrm{m/W}$. During the SBS process photons of the pump wave are scattered leading to amplification at the Stokes wave. Thereby, the gain spectrum has a Lorentzian distribution and is centered at the Stokes shift. The normalized gain distribution can be seen in Fig. 2.8.

As mentioned previously, the amplification is accompanied with a nonlinear phase change. This relation between the absorption coefficient and the refractive index in a dispersive

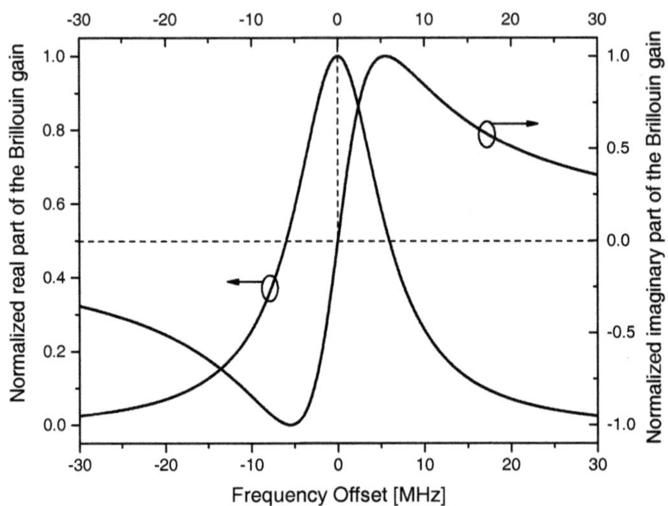

Figure 2.8.: Normalized real and imaginary parts of the Brillouin gain.

medium is a result of the Kramers-Kronig relation, that connects the real and imaginary part of the susceptibility of a medium. By taking into account the imaginary parts of Eqn. 2.13, the SBS induced phase shift for the two waves can be expressed by:

$$\frac{\partial \Phi_P^{SBS}}{\partial z} = -\frac{1}{2} g_B I_S \frac{(2\Delta\nu/\Delta\nu_B)}{1 + (2\Delta\nu/\Delta\nu_B)^2}$$

$$\frac{\partial \Phi_S^{SBS}}{\partial z} = -\frac{1}{2} g_B I_P \frac{(2\Delta\nu/\Delta\nu_B)}{1 + (2\Delta\nu/\Delta\nu_B)^2}$$

$$(2.17)$$

The normalized phase change during the SBS amplification can be seen in Fig. 2.8. It can be clearly seen that the phase change is maximum during the strongest change in the amplitude. Additionally, the phase change in the gain center is zero. Both behaviors are important for different applications.

The spectral width of the pump source directly influences the gain bandwidth of SBS. If the temporal width of the pump pulse is smaller than the phonon life time or the linewidth of the pump source is broader than the intrinsic Brillouin gain bandwidth, Eqn. 2.16 needs to be enhanced to [42]:

$$g_P = \frac{\eta_P 2\pi n^7 p_{12}^2}{c\lambda_P^2 \rho \Delta\nu_B \upsilon_a} \left(\frac{\Delta\nu_B}{\Delta\nu_B \otimes \Delta\nu_P} \right)$$

$$(2.18)$$

where $\Delta\nu_P$ is the linewidth of the pump source and $\otimes$ denotes the convolution operation between the spectral width of the pump and the Brillouin gain bandwidth. For Gaussian distributions of the pump this leads to

$$\Delta\nu_B \otimes \Delta\nu_P = \sqrt{\Delta\nu_B^2 + \Delta\nu_P^2} \tag{2.19}$$

whereas for a Lorentzian distribution it leads to

$$\Delta\nu_B \otimes \Delta\nu_P = \Delta\nu_B + \Delta\nu_P \tag{2.20}$$

As long as the $\Delta\nu_P \ll \Delta\nu_B$ there is no influence and the gain bandwidth mainly depends on the material properties rather than on the pump source. Therefore, for all narrow band laser diodes and fiber lasers simply Eqn. 2.16 can be used, as long as the power is rather low.

Beside the above mentioned influences of the pump laser linewidth on the gain coefficient, it also depends on environmental conditions like temperature or mechanical strain. This dependence can be exploited for distributed measurements of temperature and strain as will be shown in section 2.6. Additionally, the gain strongly depends on the waveguide design and material properties, as well as on the pump power, as will be shown in chapter 3.

2.4. Polarization Dependence

Since the SBS process originates from the interference between two optical waves, the efficiency is inherently dependent on the relative state of polarization (SOP) of the two waves. In optical fibers, the polarization state is usually not constant, due to small local birefringence that statistically varies along the fiber in strength and orientation. Although polarization-maintaining optical fibers have a defined and strong birefringence, the SOP is variable in propagation direction, when the light is not coupled entirely in one birefringence axis. Due to these effects, areas with parallel and orthogonal polarization states arise along the fiber for counter propagating waves. This behavior can be referred to as internal polarization mixing. All polarization occur with the same probability.

Changes of the SOP of light which passes through an anisotropic medium without polarization dependent loss, like an optical fiber, can be described mathematically as a rotation over the Poincaré sphere. The SOP of a wave that propagates in z direction can be given by the Stokes notation $S_1 = (1, s_{11}, s_{12}, s_{13})$ and for the opposite direction $S_2 = (1, s_{21}, s_{22}, s_{23})$ [43]. The mixing efficiency of two counter propagating signals with the polarizations S_P and S_S can be expressed by [44]:

$$\eta_P = \frac{1}{2}(S_1 S_2) = \frac{1}{2}\left(1 + s_{11}s_{21} + s_{12}s_{22} - s_{13}s_{23}\right) \tag{2.21}$$

For co-aligned ($\parallel$) and orthogonally ($\perp$) polarized pump and Stokes waves the mixing efficiencies become [44]:

$$\eta_{P\parallel} = \frac{1}{2}\left(1 + s_1^2 + s_2^2 - s_3^2\right) = 1 - s_3^2$$
$$\eta_{P\perp} = \frac{1}{2}\left(1 - s_1^2 - s_2^2 + s_3^2\right) = s_3^2 \tag{2.22}$$

For polarization maintaining fibers, where the SOPs of the two waves are preserved during propagation, and a linear pump polarization ($s_3 = 0$) the mixing efficiency for co-aligned and orthogonal polarizations becomes [44]:

$$\eta_{P\parallel} = 1$$
$$\eta_{P\perp} = 0 \tag{2.23}$$

Hence, by the utilization of polarization maintaining fibers it is possible to achieve a full gain or a complete vanishing of the SBS interaction by controlling the input SOPs of the light waves. For fibers with linear birefringence and a linear polarization at 45°, which means that pump power is equal in both axes, it applies independently of the polarization of the Stokes wave [45]:

$$\eta_P = \frac{1}{2} \tag{2.24}$$

Standard single mode fibers typically show a low birefringence, where random fluctuations of the waves SOP occur along the fiber. The SBS gain process for different polarizations is visualized in Fig. 2.9 [5, 44]. A linear in x-direction polarized pump wave is coupled into the fiber at the end. The fiber prevails a segmental variable birefringence. As an illustration, the segments are considered with an alternating phase delay $\Delta\varphi$ of the field components in the fast and slow axes, as well as with different orientations. This results in the drawn polarization distribution of the pump wave, as shown in Fig. 2.9(a). At the fiber input, the polarization is again linear, with an angle of -45° with respect to the x-axis. If a Stokes wave is coupled into the fiber input with the same polarization as the leaking pump wave, then it will pass through the same birefringent segments in the opposite direction, as shown in Fig. 2.9(b). Due to the opposite propagation direction, parallel polarizations arise only in some segments with identical E-fields of the pump and Stokes waves. In these cases

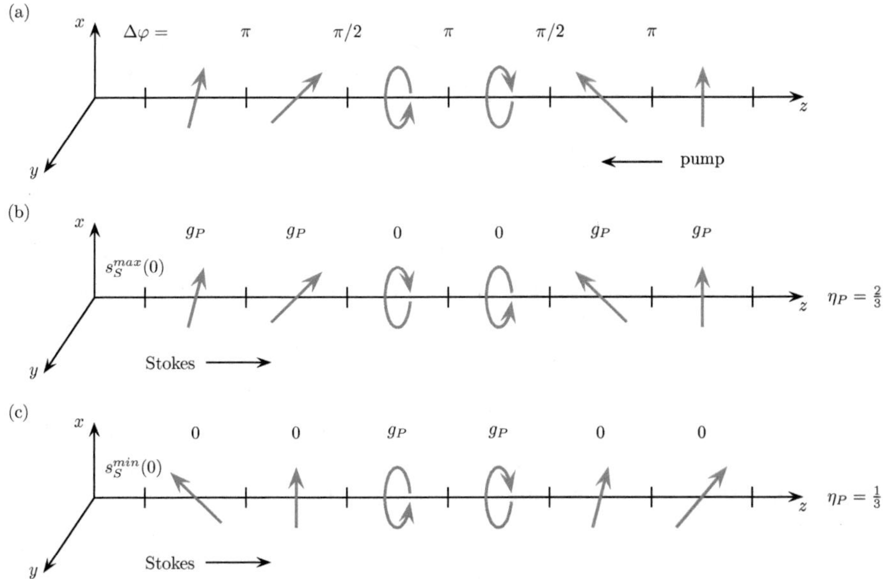

Figure 2.9.: Illustration of the polarization dependence of Brillouin gain in low birefringent fibers for an identically polarized Stokes wave (a) and an orthogonally polarized Stokes wave with respect to the pump wave.

amplification takes place, whereas in the segments with orthogonal polarization no SBS process occur. The same behavior for a Stokes wave which is coupled into the fiber at $z = 0$ with an orthogonal polarization with respect to the pump wave polarization at this point is illustrated in Fig. 2.9(c). As can be seen, the polarization mixing efficiency for co-aligned and orthogonal polarized pump and Stokes wave in low birefringence fibers are [44]:

$$\eta_{P\parallel} = \frac{2}{3}$$
$$\eta_{P\perp} = \frac{1}{3}$$

(2.25)

The equality of the birefringent properties of the pump and Stokes wave arises in the case of SBS from the fact that pump and Stokes wave have only a very small frequency difference. The chromatic dispersion of the birefringence can be neglected. Despite randomly arranged birefringent properties of the fiber, these are equal for both waves so that their relative polarization state is not completely decorrelated.

It can be derived from the previous explanation, that for a given launch SOP of the pump wave, the complex magnitude gain of the signal varies with its own input SOP, ranging between maximum and minimum values. In long birefringence fibers the maximum and minimum gain values are [46]:

$$|G_{max}|^2 = \exp\left[\frac{2}{3}g(\omega_{sig})L\right]$$

$$|G_{min}|^2 = \exp\left[\frac{1}{3}g(\omega_{sig})L\right] \tag{2.26}$$

The maximum and minimum gain values are associated with a pair of orthogonal input SOPs, which can be expressed by the Jones vectors $\hat{e}^{in}_{max}$ and $\hat{e}^{in}_{min}$. The corresponding SOPs after the amplification process can be described by $\hat{e}^{out}_{max}$ and $\hat{e}^{out}_{min}$. These input and output SOPs are governed by the SOP of the pump wave and are largely independent of ω_{sig} within the SBS bandwidth [46].

In the case of arbitrarily polarized input light and a constant polarization of the pump wave, the input signal may be decomposed on the basis of $\hat{e}^{in}_{max}$ and $\hat{e}^{in}_{min}$:

$$\vec{E}\left(\omega_{sig}, z = 0\right) = E_0\left(\omega_{sig}\right)\left(a\hat{e}^{in}_{max} + b\hat{e}^{in}_{min}\right) \tag{2.27}$$

where $E_0(\omega_{sig})$ is a scalar, frequency dependent complex magnitude and it applies $|a|^2 + |b|^2 = 1$. The corresponding output signal vector can be described by [47]:

$$\vec{E}\left(\omega_{sig}, z = L\right) = E_0\left(\omega_{sig}\right)\left[aG_{max}(\omega_{sig})\hat{e}^{out}_{max} + bG_{min}(\omega_{sig})\hat{e}^{out}_{min}\right] \tag{2.28}$$

For a sufficient strong pump $|G_{max}| \gg |G_{min}|$, and unless a is vanishing small, the output SOP of the amplified signal components is drawn towards the particular state of $\hat{e}^{out}_{max}$, which is determined by the pump polarization. This effect is referred to as polarization pulling, first shown by A. Zadok et al [46–49]. The effectiveness of the pulling is governed by the ratio $|G_{max}|\,/\,|G_{min}|$. As can be seen, the SBS introduces a difference between the output SOP of the amplified signal components and the unamplified signal components for which $G_{max}(\omega_{sig}) = G_{min}(\omega_{sig}) = 1$.

The decomposition of the Stokes wave into two orthogonal states of polarization with minimum and maximum SBS gain is also valid for the case of a saturation of the amplification by the degradation of the pumping wave [50]. As can be seen, the polarization attributes are important for the gain process of stimulated Brillouin scattering. As will be shown later in section 4.2, the proper alignment of the polarization in combination with a polarizer or polarization beam splitter can be utilized to form sharp optical filters and enhance the dynamic range of Brillouin based optical spectrum analyzers.

2.5. Threshold

An important parameter of Brillouin scattering is the threshold for the pump power in optical fibers, from where a significant part of the power is reflected [21, 22]. A number of definitions can be found in the literature [42]. In practice the measured Brillouin threshold is defined as the input optical power at which the emerging backscattered and transmitted powers are equal [51, 52].

In analytical assumptions of Brillouin scattering the threshold is also defined as the input power at which the emerging backscattered power equals the input power [22, 53]. Normally it is assumed that the pump power is significantly higher than the Stokes power and that it is not significantly depleted by the scattering process [54], but it will decrease along the fiber due to linear attenuation. The threshold directly depends on the linewidth of the pump wave. Therefore, to calculate the minimum threshold, it is assumed that this linewidth is significantly smaller than the Brillouin gain bandwidth, which is the case already for conventional telecom laser diodes with a linewidth in the lower MHz range. With the previous approximations the Brillouin threshold P_{th} for a fiber of the length L can be calculated by [22]:

$$P_{th} = 21 \frac{K_B A_{eff}}{g_p L_{eff}}, \tag{2.29}$$

where K_B is a polarization factor, A_{eff} the effective area of the fiber, g_p is the peak value of the Brillouin gain coefficient, introduced in section 2.3, and L_{eff} is the effective length of the fiber given by:

$$L_{eff} = \frac{1 - e^{-\alpha L}}{\alpha} \tag{2.30}$$

where α is the attenuation coefficient and L the length of the fiber. The polarization factor K_B depends on the polarization of the pump and Stokes waves and can be calculated by:

$$K_B = \frac{1}{\eta_{P\parallel}} \tag{2.31}$$

As shown in the previous section K_B is minimized for pump and signal waves that have the same polarization along the entire length of the fiber. It will take a value of $K_B = 1$ for polarization maintaining fibers or in general the absence of birefringence and a value of $K_B = 2$ when the pump is coupled in by 45°. For randomly polarized waves the polarization factor will be $K_B = 1.5$ [44].

Although Eqn. 2.29 is widely used in literature, there have been some adaptions of the equation due to the approximations that were made and mainly the high loss of 2 dB/km that was considered [22]. In many applications, like optical communications, already small reflected powers bother. Therefore, in [55] the threshold was defined as the input power at which the backscattered power at the fiber input is equal to 1% of the input pump power at this point. Utilizing some approximations and parameters for modern fibers with low attenuation the factor of Eqn. 2.29 was corrected from 21 to 18, also reported in [56]. The European Cooperation for Scientific and Technical research (COST) recommends a factor of 19 [57]. Finally, it has been shown that this factor is not a constant, e.g. it depends on the length of the fiber [58]. Therefore the factor converges towards the value of 16 for large fiber lengths above 20 km. In general, if the measurements will be carried out in contemporary low-loss transmission fibers, the factor 21 needs to be adapted [59]. Combining these assumptions, the threshold power for Brillouin scattering will be calculated by:

$$P_{th} = 19 \frac{K_B A_{eff}}{g_p L_{eff}}, \qquad (2.32)$$

The calculated Brillouin threshold for a standard single mode fiber, according to Eqn. 2.32 can be seen in Fig. 2.10, with an effective area A_{eff} =80 μm, K_B =2 and g_p =2x10^{-11} m/W. As can be seen the threshold decreases with increasing fiber length and approaches a limiting

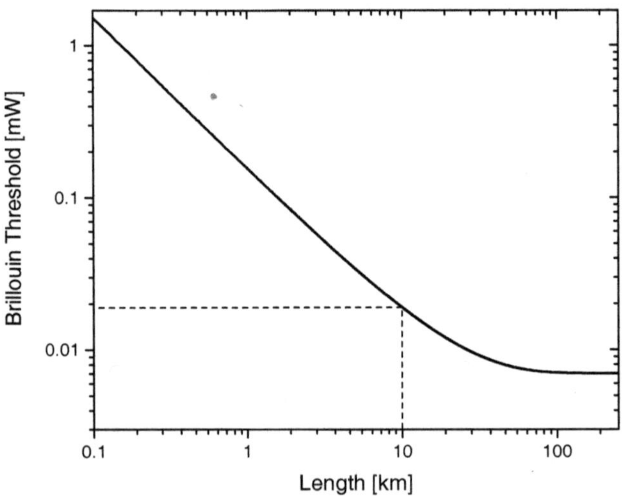

Figure 2.10.: Threshold of stimulated-amplified spontaneous Brillouin scattering.

value. This is defined by the attenuation in the fiber and the effective length, respectively. The Brillouin threshold is approximately three orders of magnitude smaller than the Raman threshold. For rather long standard single mode fibers the threshold is just 7 mW. Even a 10 km long optical fiber just requires an optical power of approximately 20 mW to generate stimulated-amplified spontaneous Brillouin scattering and a back scattering of the input power.

The setup for the measurement of the reflected and transmitted power through a fiber with respect to the input power can be seen in Fig. 2.11. The light source is a conventional distributed feedback (DFB) laser diode (LD) with a linewidth of 1 MHz and an optical output power of 10 mW. Subsequently the optical wave is amplified with an erbium doped fiber amplifier (EDFA) to provide sufficient power for the measurement. With the help of a variable optical attenuator (VOA) the desired input power can be set. The pump wave is coupled through a circulator (C) and split by a 50/50 coupler to monitor the exact pump power with an optical power meter (OPM). The other part is coupled into the fiber and the transmitted power is measured with OPM2. The reflected power is coupled out via C and measured with an optical spectrum analyzer (OSA). The losses through the coupler and circulator are measured and the real reflected power at the fiber input is calculated. During the measurement the pump is varied in order to find the threshold for the fiber under test.

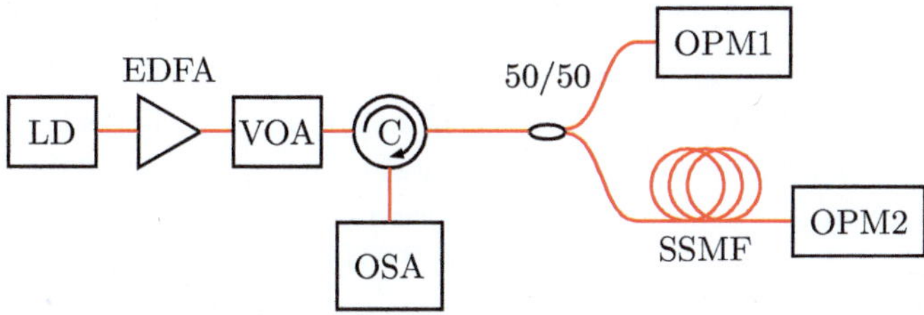

Figure 2.11.: Experimental setup for the measurement of the Brillouin threshold in an optical fiber. LD: laser diode, EDFA: erbium doped fiber amplifier, VOA: variable optical attenuator, C: circulator, SSMF: standard single mode fiber, OPM: optical power meter, OSA: optical spectrum analyzer.

During the experiment the threshold of a 10 km and a 50 km long SSMF was carried out. The measurement of the transmitted and reflected power with respect to the input power is shown in Fig. 2.12. As can be seen, the threshold for the 10 km long fiber is around an input power of 20 mW. After this point the transmitted power is saturated and a significant part of the input power is scattered back. The threshold for the 50 km fiber is with just 7 mW even lower.

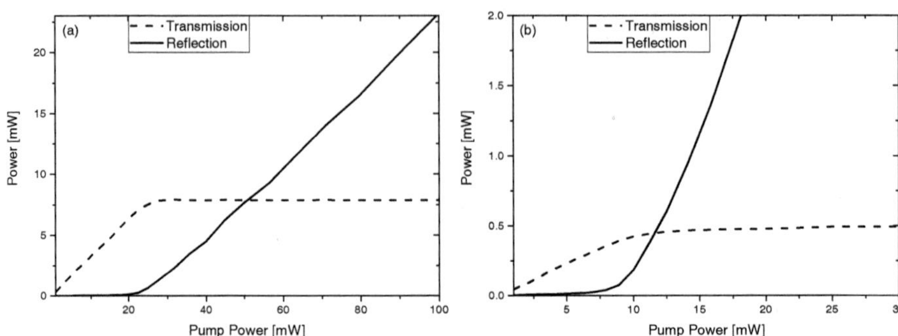

Figure 2.12.: Measurement of the transmitted and reflected power with respect to the fiber input power for a SSMF with 10 km (a) and 50 km (b) length.

As shown in the previous analysis the stimulated amplification of noise in a fiber through Brillouin scattering limits the maximum transmittable power. The maximum input power to the fiber is given through the threshold (Eqn. 2.32). The main factor is the product $P_P \cdot L_{eff}$ of the optical power and the effective fiber length. If the spectrum of the pump wave is rather small, SBS can occur easily. The stimulated amplification of noise in optical fiber due to SBS has especially drawbacks in the following applications:

- Optical communications: In optical communications very large fiber link lengths, with single path lengths in the range of 50 km and more, are used. Therefore, the effective length L_{eff} approaches the attenuation related limit $L_{eff} = 1/a \approx 20$ km. These huge lengths require just some mW of optical power to initiate the stimulated amplification of noise and therefore the back scattering of the input power, as shown in Fig. 2.10. On the one hand, this greatly limits the maximum transmittable power through the fiber, and on the other hand, the back scattered signal components can lead to instabilities of the laser itself as well as spectral distortions of the data signal, when spectral narrow band components are not transmitted, but are reflected .

- Radio over fiber: Through modulation of an optical carrier with frequencies in the microwave range it is possible to distribute high frequency signals with low attenuation over fiber optic links. This method is used in local and regional networks for cable TV (CATV), or in the supply lines of base stations to transmission units for mobile and radio links. Due to the analog transmission format, distortion and loss of quality through SBS can occur even at optical powers in the mW range.

- Narrow band fiber laser with high power: Based on double core fibers doped with rare earth metals it is possible to achieve powers in the kW range [60]. Spectral narrow band high power lasers are required for special applications, e.g. the in-phase superposition and thus coherent beam combination of several lasers to increase the overall performance with good beam quality [61]. In such systems, despite the use of short fibers in the range of a few meters, as well as fibers with large mode field diameter, the SBS is a significant performance limit [62].

2.6. Applications

Beside the mentioned disadvantages, stimulated Brillouin scattering can be used for several different kinds of applications. First, the gain which is produced by SBS in an optical fiber can be used to amplify a weak signal whose frequency is down shifted from the pump frequency by the Brillouin shift. Such **Brillouin amplifiers** were extensively studied during the 1980s [63, 64]. They are capable of providing 30 dB gain at a pump power under 10 mW [1]. However, the bandwidth of such amplifiers is rather narrow and the matching between the frequencies needs to be very accurate. For this reason, Brillouin amplifiers are not suitable for amplifying signals in fiber-optic communication systems, which commonly use erbium-doped fiber amplifiers.

The main applications of narrow band Brillouin amplifiers consists of using them as a **tunable narrowband optical filter**. Commercial optical filters with single band-reject or band-pass resonances provide wide tunability, but their bandwidth is typically in the order of tens of GHz, which is much larger than desirable for a number of applications. The spectral response for different filter types can be seen in Fig. 2.13. fiber Bragg grating (FBG) achieve bandwidths down to 5 GHz with a tuning range over 10 nm or 1.27 THz, for instance. The filter characteristic is Butterworth shaped, which leads to steep edges and a good out of band suppression. Recently, manufacturers have introduced FBG with ultra narrow bandwidths of 100 MHz, but the tuning range is limited to 20 GHz. A special type of grating based filter is a wave shaper (WS), which utilizes spatial light modulators. It can operate over a large frequency range and can be programmed individually with a bandwidth down to 10 GHz. The Gaussian filter characteristic makes them suitable as prefilter, in order to extract a small part out of a large spectrum. Furthermore, Fabry-Pérot (FP) filters can be used, which utilize the interference of light bouncing within a cavity. The filter characteristic has a Lorentzian shape. Due to the cavity, a Fabry-Pérot filter has a periodic structure. Therefore, the free spectral range (FSR) of a FP is very important. Commercially available devices achieve bandwidths of <3 GHz at 1550 nm with a FSR of up to 51 GHz. Unfortunately, the bandwidth varies in dependence of the wavelength. Therefore, custom devices are necessary to cover a large

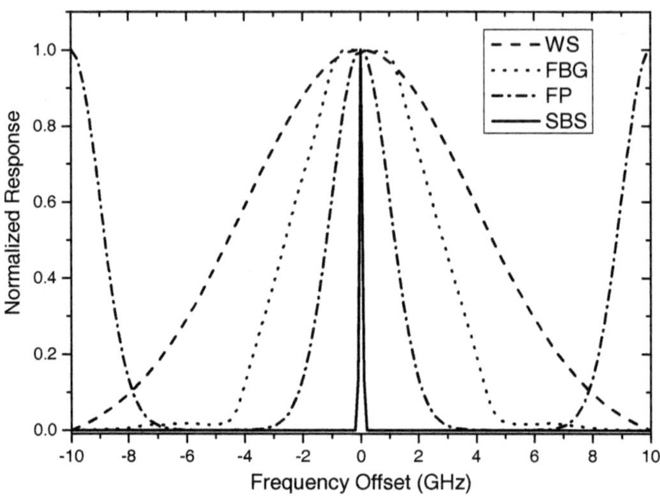

Figure 2.13.: Frequency response of different filter types

frequency range. Additional, the fabrication process for narrow bandwidths is technically challenging. As can be seen in Fig. 2.13, the bandwidth of Brillouin amplification is much narrower than all other filters. If necessary, the gain bandwidth can even be broadened artificially to form almost perfect rectangular filters [65]. These characteristics can be utilized exemplary for the channel selection in a densely packed multichannel communication system [66]. Additionally, the narrow bandwidth filter characteristics of SBS are utilized for the high resolution spectral analysis of optical signal, the storage of optical data or in general for optical signal processing of frequency combs, as will be shown later on in this work. Most of the mentioned applications would benefit significantly from a reduced gain bandwidth. Therefore, stimulated Brillouin scattering offers the possibility to form a completely tunable filter regarding bandwidth and wavelength.

On the other hand, the Brillouin gain can be used for operating a **Brillouin fiber laser**. The first realization of such a laser was reported in 1976 [67]. Most Brillouin fiber lasers use a ring cavity to avoid generation of multiple Stokes lines through cascaded SBS. Due to low resonator loss, they can have a relatively low pump threshold and a very small linewidth in the range of some kHz or even just some tens of Hz [68, 69]. An important application of CW Brillouin lasers consists of using them as a sensitive laser gyroscope [70].

Besides the narrow band optical filtering, SBS can be utilized for **distributed sensing** of temperature and strain changes over relatively long distances, first introduced in the early

1990s [71–74]. The operation principle is based on the dependency of the Brillouin frequency shift on the effective refractive index of the fiber mode, which changes whenever the refractive index of silica changes in response to local environmental variations. Both temperature and strain can change the refractive index of silica. By monitoring changes in the Brillouin shift along fiber length, it is possible to map out the distribution of temperature or strain over long distances over which the SBS signal can be detected with a good signal-to-noise ratio. The basic idea has been implemented in several experiments to demonstrate distributed sensing over distances as long as 32 km. Therefore, a tunable continuous wave probe laser and a pulsed pump laser are injected at the opposite ends of a fiber. The frequency difference between the two lasers could be set to a particular value corresponding to the Brillouin frequency of the optical fibers, and the continuous wave (CW) probe would experience gain varying along the fiber. The gain as a function of position along the fiber could thus be determined by the time dependence of the detected CW light. By measuring the time dependent CW signal over a wide range of frequency differences between the pump and probe, the Brillouin frequency at each fiber location could be determined. This allowed mapping the strain or temperature distribution along the entire fiber length. The performance of distributed sensors based on Brillouin scattering has been improved significantly over the years. The highest spatial resolution of 1.2 cm was achieved over 20 m sensing length by utilizing Brillouin dynamic gratings [75] and the longest reported sensing length is 150 km with 2 m spatial resolution and 1 °C temperature resolution [76].

Another important field of applications are **microwave photonic filters**. They represent a great advantage to overcome the limitations of microwave filters implemented in the electrical domain, i.e. difficulties in tuning, reconfiguration, filter shape and its bandwidth as well as electromagnetic interference [77]. In general, microwave photonic filter (MPF) can overcome these limitations and are predestined for microwave signal processing with applications in radar, radio over fiber and mobile communication. Recently, widely tunable single pass-band MPF with high Q-factors have been demonstrated by exploiting the narrow gain distribution of SBS [65, 78, 79]. Additionally, the Brillouin loss can be utilized in the same manner as tunable notch filter [80]. The filtering properties, including the out of band rejection and the bandwidth, directly depend on the pump power and pump wave configuration [81]. MPF with complex coefficients can be realized with a dual pump wave system, enabling independent manipulation of the magnitude and phase of an optical carrier [82]. The utilization of frequency combs as pump source for the SBS leads to reconfigurable passband filters with a rectangular shape, tunable bandwidth and high selectivity [83]. But, the minimum filter bandwidth is given by the Brillouin bandwidth itself.

The bandwidth reduction of stimulated Brillouin scattering, as presented in the next chapter of this work, can significantly enhance the performance of these applications or enable completely new fields, as shown later on.

3

Brillouin Bandwidth Reduction

Many parameters are able to modify the properties of the Brillouin gain spectra. The first one, as often in physics, is the temperature. In general, the effects of temperature changes around ambient temperatures on the Brillouin properties are well known in the field of distributed sensing and have been studied intensively [84–86]. Accordingly, the Brillouin frequency shift has a linear dependence on the temperature in the range of -40–100°C with an increasing frequency shift of 1.36 MHz/°C [86]. An interesting aspect is the decrease of the Brillouin linewidth with increasing temperature in the mentioned temperature range [86]. More significant effects on the linewidth occur at extreme temperature values. The Brillouin linewidth $\Delta\nu_B$ in dependence of the temperature can be seen in Fig. 3.1. The curves show an approximation based on the measurement results presented in [87] and [88].

The measurements were carried out in SSMF at atmospheric pressure and in the undepleted pump regime where $G < 2$. The temperature ranges from cryogenic achieved in liquid helium

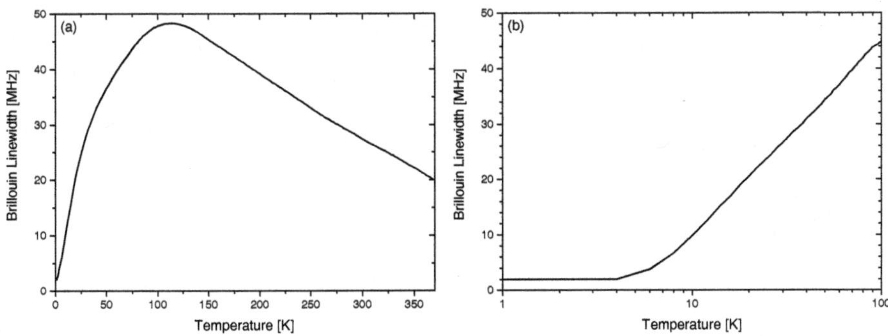

Figure 3.1.: Dependence of the Brillouin linewidth on the temperature. The graphs are based on the measurements carried out in [87] and [88].

over liquid nitrogen and ambient temperatures up to heating the fiber in an oven. As can be seen there is a maximum linewidth occurring at the temperature of 110 K, based on the phonon absorption in the material [89]. The linewidth decrease for higher temperatures possibly represents the upper tail of the absorption peak. Below the absorption maximum the Brillouin linewidth decreases rapidly. This behavior originates in tunneling effects that occur at very low temperatures. With increasing temperature the defect states can reach on of the two sides by thermic agitation [88, 90]. For temperatures below 4 K the Brillouin linewidth reaches a plateau. In total, minimum linewidths of 2–3 MHz have been demonstrated at very low temperatures of 4 K, as can be seen in Fig. 3.1(b) [87, 91].

Interestingly, the Brillouin gain peak value exactly compensates the narrowing of the Brillouin linewidth resulting in a constant value of the product $g_p(T)\Delta\nu(T)$ [86, 87]. For very low temperatures this leads to an enormous gain. However, the Brillouin linewidth is not sufficiently reduced in the ambient temperature range. Cryogenic temperatures are critical for the most applications and standard telecom fibers can rarely handle high temperatures. Therefore, other methods for the Brillouin gain linewidth reduction need to be applied.

Other physical parameters that effect the Brillouin linewidth are the fiber geometry itself and possible dopants of the fiber. The refractive index profiles for different fiber types can be seen in Fig. 3.2 [92]. The fiber properties are designed for different tasks, e.g. a TrueWave fiber (TW) and an AllWave fiber (AW) have no water peaks left and dispersion management is possible with a dispersion compensating fiber (DCF) and a dispersion shifted fiber (DSF). Typical parameters such as Germanium concentration and diameter can be seen in Tab. 3.1. Additionally, the Brillouin linewidth $\Delta\nu_B$ of the spontaneous and stimulated scattering, as well as, the Brillouin shift are mentioned.

As can be seen, the linewidth of stimulated Brillouin scattering varies for the different fiber types. Especially, in dispersion-shifted fibers linewidths of 7.2 MHz could be achieved [56].

Table 3.1.: Properties of different fiber types. AW: AllWave, TW: TrueWave, DCF: dispersion compensating fiber, DSF: dispersion shifted fiber [92].

Fiber Type	Ge %	D_{eff} μm	$\Delta\nu_B^{spon}$ MHz	$\Delta\nu_B^{stim}$ MHz	ν_B GHz
AW	3.6	9.37	86	11.4	10.81
TW	5.7	7.70	88	12.4	10.85
DCF	18	2.50	92	12.6	9.75
DSF	6.2	7.94	37	7.2	10.55

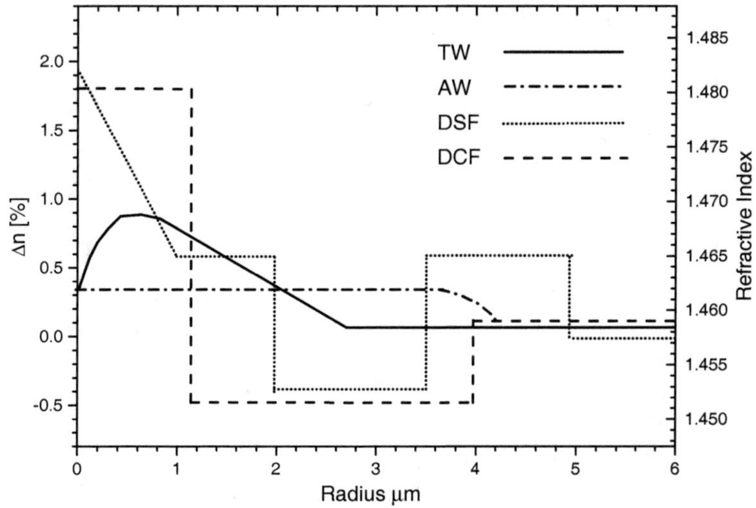

Figure 3.2.: Refractive index profiles for different fiber types.

This behavior is mainly based on the different mode field diameters. Additionally the doping of the fibers plays an essential role. It has been shown that different dopants affect directly the acoustic velocity and the refractive index of silica, leading to a shift of the Brillouin gain center [93, 94]. In order to design regions with different refractive indices mainly Germanium is used as dopant. This leads to a shift of the Brillouin frequency by -68 MHz/mol% [95]. At the same time the Brillouin linewidth broadens due to the larger acoustic attenuation of GeO_2 by 1.3 MHz/mol% [95]. In case of the DCF, the combination of higher Germanium doping and concurrently a smaller mode field diameter still leads to SBS linewidths in the range of 12 MHz. However, with changing the Germanium concentration of a fiber there is no possibility to reduce the gain bandwidth significantly. In general, the manufacturing of special fiber geometries and doped fibers is expansive and most applications will utilize cheap standard single mode fibers.

The last parameter to change might be the strain of the fiber. It is well known from distributed sensing that the strain has an effect on the Brillouin frequency shift. But, several measurements have revealed that there is no influence on the Brillouin linewidth [86]. In order to reduce the Brillouin linewidth at room temperatures without changing the geometry or physical parameters of the used fiber, the pump power can be taken into account.

3.1. Pump Power Dependence

Theoretical predictions have shown that the Brillouin linewidth varies considerably with the Brillouin gain and hence the pump power [96]. Boyd et al. have shown theoretically, by taking the Fourier transform of the autocorrelation function of the Stokes wave, that the Brillouin gain spectrum or the spectral density of the Stokes wave can be written as [97]:

$$S\left(\omega\right) = \frac{8\pi\hbar\omega_S\left(\tilde{n}+1\right)}{ncA\Gamma}\left[\exp\left(\frac{G\left(\Gamma/2\right)^2}{\omega^2+\left(\Gamma/2\right)^2}\right)-1\right] \tag{3.1}$$

where Γ is the phonon decay rate, $\tilde{n}$ is the mean number of phonons per mode of acoustic field, A is the effective cross section area of the gain medium and the single pass gain $G = g_B I_L(0)L$, where g_B is the Brillouin gain factor, I_L is the incident pump intensity and L is the gain medium length. As can be seen, the spectral shape of the Brillouin gain depends on the single pass gain and on the phonon decay rate, which is the reciprocal of the acoustic phonon life time in the medium. Within the low gain limit (G≪1) the Brillouin gain will be Lorentzian shaped and can be written as [97]:

$$S\left(\omega\right) = \frac{8\pi\hbar\omega_S\left(\tilde{n}+1\right)\Gamma G}{ncA}\left[\frac{1}{\omega^2+\left(\Gamma/2\right)^2}\right] \tag{3.2}$$

which is characteristic for the spontaneous Brillouin scattering. The full width at half maximum bandwidth will be given by [98]:

$$\Delta\nu_B = \Gamma = \alpha_a v_a \tag{3.3}$$

where α_a is the attenuation and v_a the speed of the acoustic wave in the medium. With increasing pump power the gain will be increased to $G \gg 1$ and the shape of the Brillouin gain becomes Gaussian. Therefore, Eqn. 3.2 can be written as [97]:

$$S\left(\omega\right) = \frac{8\pi\hbar\omega_S\left(\tilde{n}+1\right)}{ncA\Gamma}e^G\exp\left[-\left(4G\omega^2/\Gamma^2\right)\right] \tag{3.4}$$

The spectral shape of the Brillouin gain in this gain region will be given by:

$$\Delta\nu_B = \Gamma\left(\frac{ln2}{G}\right)^{1/2} \tag{3.5}$$

The foregoing indicates that the Brillouin bandwidth is dependent on G. As the Brillouin pump power is increased from below the threshold , the gain increases and the spectral shape of the gain undergoes a change from Lorentzian to Gaussian. The narrowing of the spectrum with increasing values of G is termed gain narrowing. The calculated and measured SBS linewidth in dependence of the pump power for a 20 km long AllWave fiber can be seen in Fig. 3.3(a). The Brillouin threshold was measured to be 10.5 dBm in accordance with Eqn. 2.32. As can be seen, below the threshold, the gain bandwidth is, as expected, much broader. Above the threshold, and clearly in the stimulated process, it approaches a constant value around roughly above 10 MHz. From the calculation it could be assumed that the linewidth eventually goes to zero for very high gain, which could be obtained by increasing the pump power or the interaction length of the fiber. On the one hand, the effective length will approach a fixed limit according to Eqn. 2.30. Additionally, as with every amplification process, the SBS shows saturation. Thus, the maximum achievable gain is restricted to $g_{0max} \approx 19$ [99, 100]. Additionally, a zero value of the Brillouin linewidth corresponds to an infinite phonon life time, which is insubstantial. Hence, the minimum Brillouin linewidth for high pump powers is around 10 MHz.

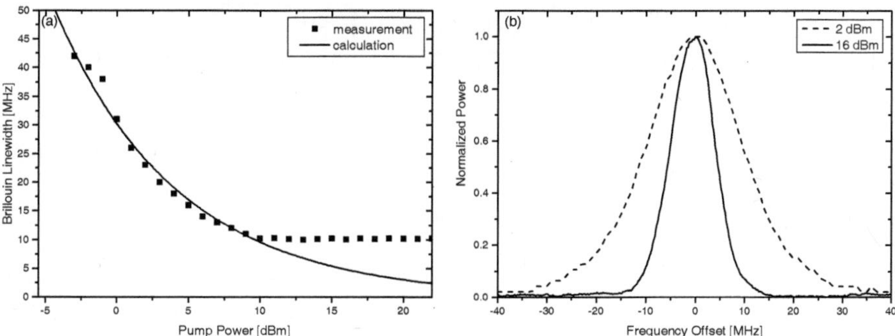

Figure 3.3.: (a) Comparison of the calculated Brillouin linewidth versus measured values and (b) exemplary measured gain profiles.

The measured gain profiles for two exemplary chosen pump powers of 2 and 16 dBm can be seen in Fig. 3.3(b). The measured Brillouin linewidth for a pump power of 2 dBm is 23 MHz and for a pump power of 16 dBm it is reduced to 10.3 MHz. Further measurements of different fiber types can be found in [92]. Unfortunately, the increase of the SBS related noise, accompanied with high pump powers, makes this approach impractical for many applications. As can be seen, the linewidth clearly approaches a limit defined by the phonon decay rate in the material. Therefore, different approaches need to be found to reduce the Brillouin linewidth even further.

3.2. Multi Stage System

As shown previously, the Brillouin gain bandwidth will be narrowed by increasing pump powers. But, it approaches a limit of $\approx 10\,\mathrm{MHz}$ at room temperature in standard fibers and could be as low as $7.2\,\mathrm{MHz}$ in dispersion shifted fibers. The first method for the gain bandwidth narrowing of stimulated Brillouin scattering is based on cascading of several Brillouin amplifiers [101]. Despite the bandwidth narrowing, this approach benefits from a variable gain. The basic principle of a SBS amplifier with n equal stages can be seen in Fig. 3.4. In every stage the signal is coupled into the SBS block, consisting of a fiber, a circulator and a counter propagating pump wave, and will be amplified. Afterwards, the amplified signal is attenuated at the end of every stage in order to provide the same initial condition for each stage and to avoid saturation of the amplification process.

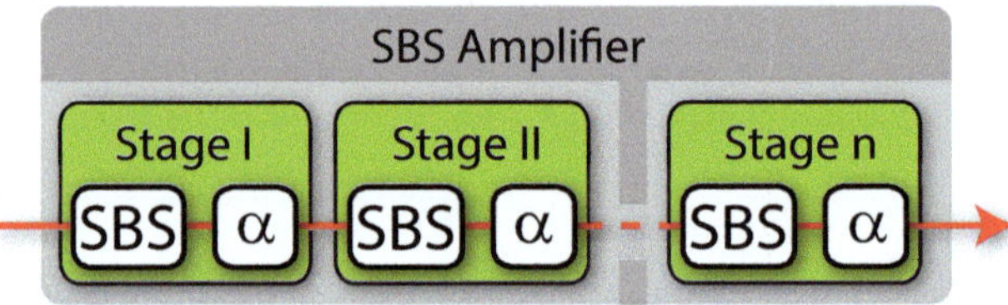

Figure 3.4.: Basic concept of a SBS amplifier with multiple equal stages. Each SBS block consists of a fiber, a pump wave and a circulator. The α block symbolizes an attenuator.

The gain for such a multi stage system can be written as [101]:

$$G_{ges} = e^{ng-n\alpha} = exp\left(\frac{ng_p(\frac{\Gamma}{2})^2}{(\omega-\omega_0)^2 + (\frac{\Gamma}{2})^2} - n\alpha\right) \tag{3.6}$$

with α as the attenuation of the attenuator block. The spectral width of this n-stage SBS amplifier can be written as:

$$\Delta\nu_B = \Gamma\sqrt{\frac{ln2}{ng_p - ln2}} \tag{3.7}$$

As can be seen, the bandwidth of the SBS decreases with the square root of the number of stages $\Delta\nu_B \sim n^{-0.5}$. Exemplary, for $n = 2$ the minimum Brillouin gain bandwidth is around $7\,\mathrm{MHz}$ and for $n = 4$ it would be $5\,\mathrm{MHz}$. Though, the gain bandwidth also depends on the line center gain g_0 and therefore on the fiber parameters. As shown previously, different temperatures, strains or doping of the fiber lead to a different value for the SBS shift of the respective fiber. Therefore, to guarantee a uniform amplification of the signal, all the used fibers need to have the same Brillouin frequency shift. Otherwise several single gains or one broadened gain might occur in the fiber and the gain bandwidth won't be narrowed.

The experimental setup for the investigation of the theoretical assumptions can be seen in Fig. 3.5. The signal wave for the multi stage amplifier is generated by a DFB laser diode (LD1) and is subsequently processed with a Mach-Zehnder modulator (MZM). The modulator is driven by f_{scan} in order to shift one of the generated sidebands through the Brillouin gain with a fine resolution. Afterwards the signal is coupled into the fiber of the first stage. The pump wave for the whole system is generated by LD2. To provide sufficient pump power for the first stage, the pump wave is amplified by an EDFA. Thereby the output power of EDFA1 was set to 24 dBm. Afterwards it is coupled into a standard single mode fiber via a circulator (C). The fibers in the different stages have approximately the same length and a Brillouin shift of $f_{SBS} = 10.873$ GHz at room temperature. The measured value for the single pass gain of the different stages is $G \approx 23$ dB.

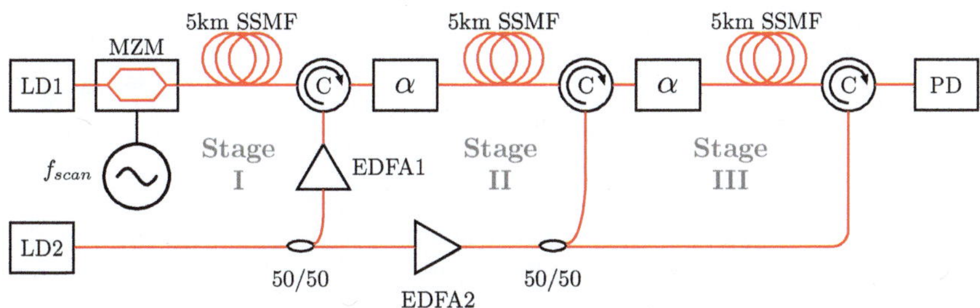

Figure 3.5.: Experimental setup for the multi-stage Brillouin system. LD: laser diode; MZM: Mach-Zehnder modulator; EDFA: Erbium-doped fiber amplifier; Att: attenuator; SSMF: standard single mode fiber; PD: photo diode

The amplified signal is coupled out of the fiber via the circulator and is attenuated by 35 dB before entering the following stage. This leads to a signal power of around -34 dBm for stage 2 and around -33 dBm for stage 3. Due to a lack of equipment the pump powers for stage 2 and 3 are both provided by EDFA2 with an output power of 27 dBm. However, it can provide sufficient pump power for the experiment. After stage 3 the signal is detected with a photo diode (PD). The power of the detected signal is recorded with respect to the frequency change of f_{scan}. Therefore, the signal is shifted through the gain and the spectral width of the Brillouin gain can be measured.

Within the measurement the number of stages was limited to 3 stages, since all other available fibers have had a different Brillouin shift. Within the experiment, the fibers had a length of 5 km. The length for each segment can be varied in order to adapt the SBS gain to the available pump power, as long as all fiber segments have the same Brillouin shift. The measured and normalized gain spectra after the different stages can be seen in Fig. 3.6(a). For the first stage we were able to achieve a single pass Brillouin gain bandwidth of 10.3 MHz. The dashed curve indicates a 7 MHz bandwidth after the second stage and the solid curve a bandwidth of 5.8 MHz after the third stage. A comparison of the measured values and the calculated ones is shown in the inset in Fig. 3.6(b). As can be seen the measurement fits very well with the predicted theoretical values.

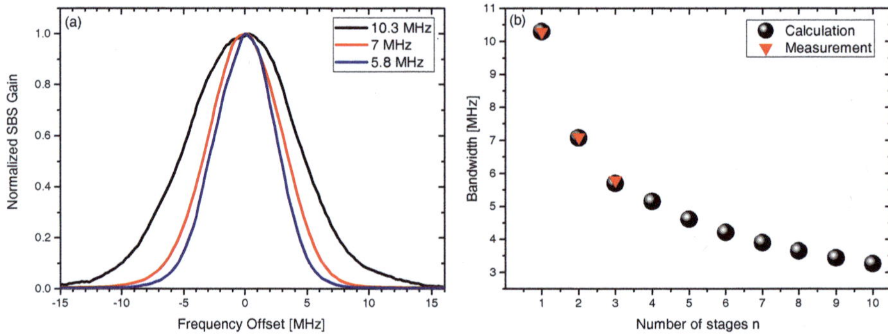

Figure 3.6.: (a)Measured and normalized SBS gain for one (black), two (red) and three (blue) SBS amplifier stages. (b) shows the measured (triangles) and calculated (spheres) Brillouin gain bandwidths vs. the number of stages

Theoretically, the system can be reduced to one stage and utilize an external feedback [102]. In general, that would lead to a Brillouin laser with possible linewidths in the lower kHz range. Therefore, it is rather difficult to couple in the signal or data in this type of feedback amplifier and prevent the whole system from lasing. First experiments in this regard have been carried out during the investigation of the storage time enhancement of Quasi-Light-Storage and can be found in section 5.5.3. Despite the feedback system, many applications which exploit this narrow filter function of the SBS, like optical filters for millimeter and THz-waves and ultra high resolution spectroscopy can benefit from the reduced bandwidth [103].

3.3. Frequency Domain Aperture

Another possibility for the gain bandwidth reduction is the utilization of a frequency domain aperture [104]. Thereby, the saturation characteristics of the SBS gain are utilized to reduce the bandwidth. A deeper analysis and theoretical calculations can be found in [105]. The amplification of the SBS process depends directly on the input power of the signal. If the signal intensity is much smaller than the pump intensity, the pump is assumed to be independent of the signal [4]. If the signal power is coming into the order of magnitude of the pump, the pump power is depleted, which limits the gain and leads to a saturation of the amplification process, where further amplification of the signal power becomes impossible.

Fig. 3.7 shows the input signal to the SBS system at two different levels. The dotted black line shows the low signal under test power level and the solid black line shows the signal superposed with the aperture. This higher aperture power level will saturate the amplification process. For the SBS interaction, this results in a small gain for the aperture components and a larger gain for the signal components inside the gap. If the gap in the saturating signal spectrum is smaller than the natural SBS bandwidth (dashed black line), the edges of the gain are suppressed due to the saturation and the gain bandwidth is narrowed to the spectral

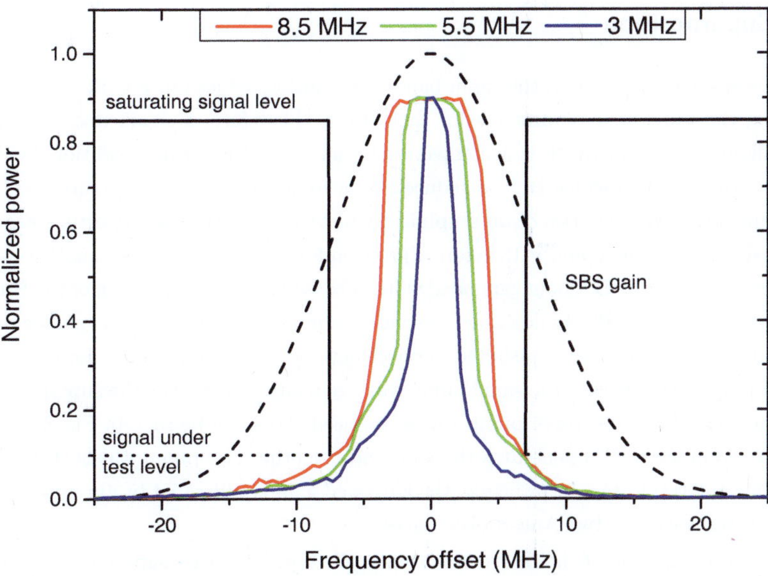

Figure 3.7.: The fiber input signal as a superposition of the signal under test (dashed black line) and the aperture signal (solid black line) for an aperture bandwidth of 15 MHz. The dashed line shows the SBS gain that would occur for a small signal power without an additional aperture. The colored lines show measured gain distributions with a saturating aperture at different gap widths [104].

width of the gap.

The measured gain distributions with different widths are shown in Fig. 3.7. The red, green and blue lines show the SBS gain distributions by superposing the saturating aperture, with achieved bandwidths of 8.5, 5.5, and 3 MHz, respectively [104]. This equals a maximum reduction of the bandwidth down to 15% of the natural one. However, further reduction of the gain bandwidth into the kHz range should be achieved through an appropriate generation of the aperture in the electrical domain. In general, the generation of narrow gaps with steep edges or a perfectly rectangular shape is experimentally challenging. Additionally, the used aperture has to be extracted from the signal and is therefore a problem for some applications, e.g. the data processing, where the aperture would destroy the information in the signal. Also for very narrow gaps the short-time laser drift needs to be taken into account and stabilized. Nevertheless, the great advantage of this method is, that the gain in the center is not reduced, which promises a good contrast or a large dynamic range for specific applications.

3.4. Superposition

It has been shown previously, that gain bandwidth can be reduced in a multi stage system, requiring the same fiber parameters for each stage. Therefore, a significant reduction of the Brillouin gain bandwidth is accompanied by an enormous setup. Additionally, it has been shown that bandwidth can be reduced by a frequency domain aperture, exploiting the saturation characteristics of an amplifier. Unfortunately, the used aperture has to be extracted from the signal and is therefore a problem for some applications. Another method for the reduction of the Brillouin gain bandwidth is based on the superposition of the Brillouin gain with two losses [106]. As described previously, a pump wave generates a gain for counter propagating waves if they are up-shifted in wavelength by the Brillouin shift. For downshifted counter-propagating waves, the same pump wave generates a loss. For the superposition of the gain with two losses, three pump waves are needed, as can be seen in Fig. 3.8. There, the first pump wave (P1) generates the gain and the other two pump waves (P2, P3) are generating the two losses. In principle, the narrowed gain is generated by the superposition of one Stokes wave and two Anti-Stokes waves.

The gain distribution of the SBS mechanism in the fiber for a gain superimposed with two losses can be written as [107, 108]:

$$G_{ges} = \frac{g_0 \Delta\nu_B^2}{\Delta\nu^2 + \Delta\nu_B^2} - \frac{g_1 \Delta\nu_B^2}{(\Delta\nu + \delta)^2 + \Delta\nu_B^2} - \frac{g_1 \Delta\nu_B^2}{(\Delta\nu - \delta)^2 + \Delta\nu_B^2} \tag{3.8}$$

where g_0 is the maximum gain, g_1 is the maximum loss, $\Delta\nu_B$ is the half width at half

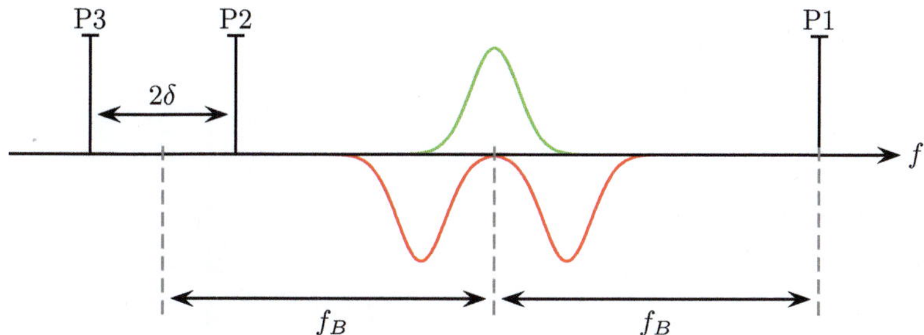

Figure 3.8.: Spectral distribution of the pump waves for the superposition of the Brillouin gain with two losses.

maximum linewidth of the Brillouin gain, $\Delta\nu$ is a detuning frequency given by $(\nu - \nu_B)$, and δ is the separation between the losses. By normalizing Eqn. 3.8 with $\Omega = \frac{\Delta\nu}{\Delta\nu_B}$; $m = \frac{g_1}{g_0}$ and $d = \frac{\delta}{\Delta\nu_B}$ it follows:

$$G_{ges} = g_0 \left(\frac{1}{\Omega^2 + 1} - \frac{m}{(\Omega + d)^2 + 1} - \frac{m}{(\Omega - d)^2 + 1} \right) \tag{3.9}$$

As can be seen, the overall gain bandwidth can be narrowed by varying the distance d of the losses and the ratio m between gain and losses. The theoretical calculation for the gain bandwidth reduction can be seen in Fig. 3.9. The reduced gain is normalized to the natural gain and the gain/loss ratio m is varied from 0.1 to 1 in steps of 0.05. As can be seen, the gain bandwidth can be narrowed to 50% (point 1, m=0.45, d=0.6), 20% (point 2, m=0.55, d=0.4) or even below by the superposition of the gain with two losses. If the separation of the losses is very low and the power of the losses is much stronger than the power of the gain, the overall gain vanishes and the signal will be attenuated. Therefore, the graph shows a bandwidth ratio of zero.

Since the gain is overlapped by two losses, this narrowing comes at the cost of a lower gain. The theoretical calculations are shown in Fig. 3.10. It can be clearly seen, that for the same points as before (point 1, m=0.45, d=0.6; point 2, m=0.55, d=0.4) the overall gain is reduced. For a bandwidth reduction of 50% the remaining gain has only 33% of the natural gain. The reduction down to 20% leads to a remaining gain of just 5%. The reduced gain can be an advantage for some applications. Since Brillouin scattering is an amplification process, it adds noise to the signal and decreases the signal to noise ratio (SNR). In order to achieve a Brillouin linewidth of 11 MHz in an AllWave fiber just by conventional SBS, very high pump powers beyond the threshold are needed, as explained previously. Consequently, a high

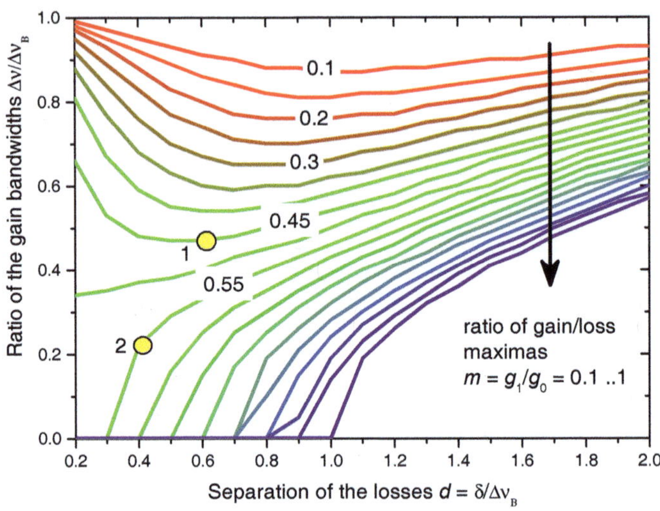

Figure 3.9.: Calculated gain bandwidths with respect to the natural bandwidth as a function of frequency separation for several gain/loss ratios.

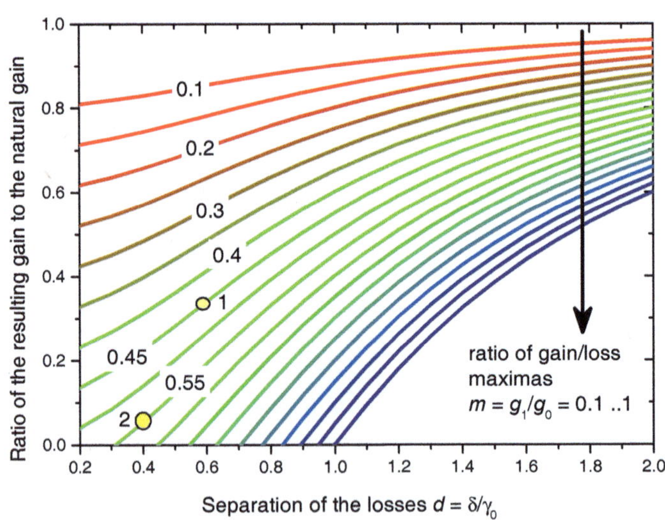

Figure 3.10.: Calculated gain with respect to the pure gain in dependence of the frequency separation between the losses for several gain/loss ratios.

amount of noise is added to the signal. Additionally, the SBS process is in the depleted pump regime and most of the theoretical assumptions do not hold anymore. For the superposition method the power of the overall gain and the noise are smaller for the reduced bandwidth. However, in some applications this reduced gain can decrease the SNR. But it is also possible to compensate the reduction of the gain by the two losses with higher pump powers. In the above equations only the ratio between the gain and losses is important. However, since the wave which generates the losses is generating two gains as well, the maximum achievable pump power for the losses is defined by the threshold of SBS in the fiber.

The spectral representation of two reduced gains in comparison to a natural gain can be seen in Fig. 3.11. In this case a negative gain means a suppression of the spectral components of the counter propagating signal which fall in this area, due to the characteristics of the losses. For applications which are based on the narrow band extraction of frequency components, this is an additional advantage of the method since it can enhance the SNR or the dynamic range.

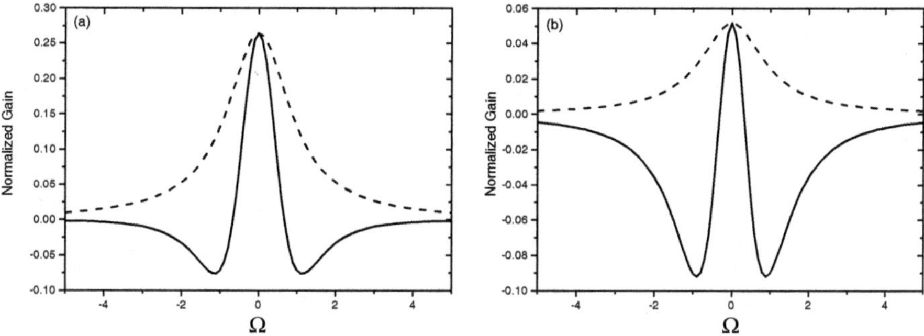

Figure 3.11.: Spectral distribution of the reduced single gain (dashed) and the superimposed gain with two losses (solid). Subfigure (a) corresponds to point 1 and (b) to point 2.

The theoretical predictions are confirmed with an experiment. The experimental setup can be seen in Fig. 3.12. In order to stabilize the relation between gain and the two losses, both are produced by the LD on the right hand side. The LD has a linewidth of 1 MHz at a wavelength of 1550 nm. The MZM4 is driven in the suppressed carrier regime with a sine wave at the frequency of the Brillouin shift f_B, which is 10.855 GHz for the used fiber. The two sidebands have a distance of twice the Brillouin shift. Afterwards the signal is split via a 3dB coupler. In the upper path the sideband with the higher wavelength is filtered out with a FBG, so that the lower sideband can be used as the gain pump, as illustrated in Fig. 3.8. In the lower path the sideband with the higher wavelength is selected and utilized in

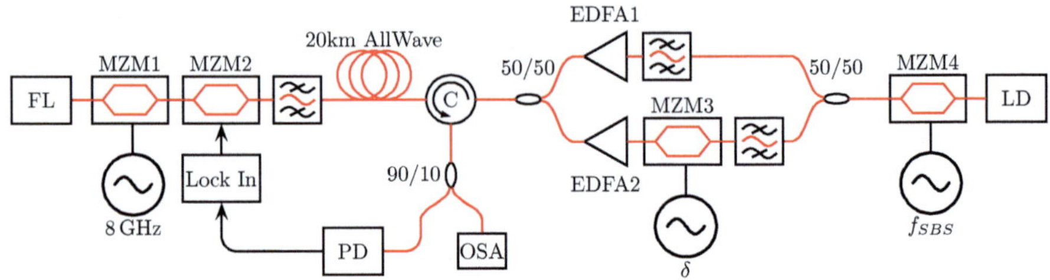

Figure 3.12.: Experimental Setup for the bandwidth measurement of the gain superimposed with two losses. FL: fiber laser, MZM: Mach-Zehnder modulator, PD: photo diode, OSA: optical spectrum analyzer, LD: laser diode, C: circulator.

combination with MZM3 to produce the two losses. MZM3 is driven with a sine wave with a frequency corresponding to the parameter δ in Eqn. 3.8. Afterwards both pump waves, for the gain and the two losses, are independently amplified by an EDFA and combined via a coupler.

On the left hand side the setup for scanning the reduced gain can be seen. Therefore, a narrow band signal is shifted through the superposition of the gain with the two losses. A fiber laser (FL) with a line width of 1 kHz at a wavelength of 1550 nm with an output power of 13 dBm is modulated via MZM1. In order to adapt the wavelength of the fiber laser to the LD and to perform a wavelength scan. The modulation frequency is 8 GHz and is swept by ±100 MHz in 0.5 MHz steps during the scanning process. With MZM2 the reference frequency for the lock in amplifier is modulated to the signal. Subsequently, the lower frequencies are filtered out with a FBG. The remaining signal falls in the spectral region where the reduced gain is generated. The fiber laser is just used for measuring the spectrum of the resulting signal. Due to the modulation, the power of the fiber laser is too low to create a stimulated Stokes or Anti-Stokes wave in the fiber. The input power into the fiber is 10 dBm.

As propagation medium a 20 km AllWave fiber ($L_{eff} = 13.38km$, $\alpha = 0.19dB/km$, $A_{eff} = 86\mu m^2$) was used. The pump waves are coupled into the fiber via a circulator. The backscattered wave is split with a 90/10 coupler and then detected with an OSA and a PD. The correct position of all waves can be monitored by the OSA. The electrical signal from the PD is measured by the Lock In amplifier. For every frequency step the corresponding value for the amplitude is recorded.

The natural gain bandwidth in the used AllWave fiber for the undepleted pump power regime was measured to be 20 MHz. For a superposition of gain and losses with the parameters $m = 0.45$ and $d = 0.6$, corresponding to point 1 from the theoretical part, the measured result can be seen in Fig. 3.13(a). During the measurement the output power of the EDFA1

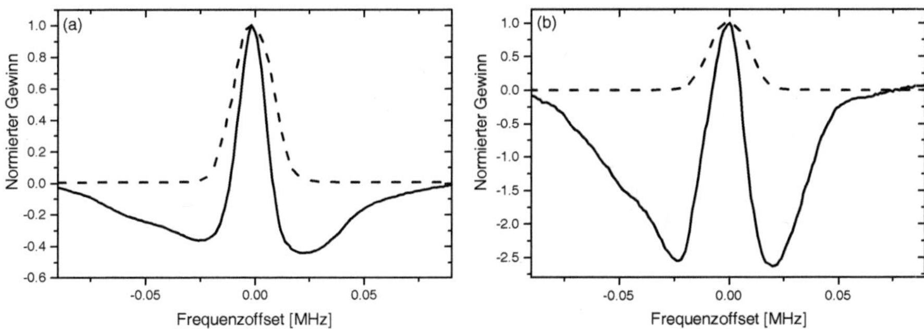

Figure 3.13.: Measurement results for a single Brillouin gain (dashed) in comparison to the superposed gain (solid) for point 1 (a) and point 2 (b).

was 27 dBm for the Brillouin gain and at EDFA2 23.5 dBm for the Brillouin losses, resulting in a gain/loss ratio of $m = 0.45$. The losses are separated by 12 MHz which results in $d = 0.6$. The measured gain bandwidth is 10 MHZ, which agrees well with the predicted reduction down to 50%.

The measurement of the parameters for point 2, $m = 0.55$ and $d = 0.4$, can be seen in Fig. 3.13(b). Therefore the pump power of the losses was increased to 24.4 dBm and the losses are separated by 8 MHz. The reduced gain bandwidth is measured to be 8 MHz, which equals a reduction down to 40% of the natural gain bandwidth. As can be seen the results differ from the simulation. Within the measurement the gain bandwidth is not reduced as much as in the simulation. The measured 8 MHz equal a reduction of 40% of the Brillouin gain bandwidth and not 20% as stated in the simulation. A reason could be that the theory of Eqn. 3.8 and Eqn. 3.9 holds just for small signals. In order to show the suppression by the losses in Fig. 3.13(b), the signal for scanning the spectral distribution of the superimposed gain had to be with 10 dBm quite high.

A further bandwidth reduction is possible by a decrease of the distance of the losses. In Fig. 3.14 the results for $m = 0.55$ and $d = 0.3$ can be seen. Therefore, the pump powers for gain and loss were kept and the frequency difference between the two losses was decreased to 6 MHz. For these values, a reduction of the bandwidth down to 3.4 MHz was achieved. This equals to a reduction of 17% of the original Brillouin gain bandwidth. Unfortunately, the residual gain is just a small fraction of $\approx 1\%$ of the original gain. A further decrease of the distance between the losses would lead to a vanishing of the residual gain. However, this small gain is not a disadvantage since the adjacent spectral components are strongly suppressed by the losses, resulting in a steep filter. Additionally, the previously measured

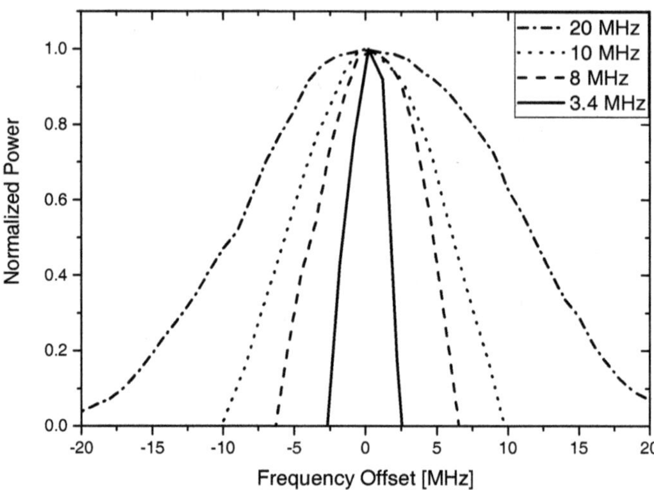

Figure 3.14.: Different measurements for the gain bandwidth reduction by superposition of a SBS gain with two spectral shifted SBS losses.

Brillouin gain bandwidths are represented in Fig. 3.14 as well.

As have been demonstrated, it is possible to reduce the Brillouin gain bandwidth significantly with several methods. The following chapters will introduce applications which utilize stimulated Brillouin scattering as narrow band filter or amplifier. All of them will clearly benefit from the bandwidth reduction.

4

Optical Spectrum Analysis

The acquisition of optical power spectral densities is among the most fundamental and widely practiced procedures in optical test and measurement. Optical spectrum analyzers are routinely employed in testing laser and LED light sources for spectral purity and power distribution, monitoring of optical communication networks, e.g. optical signal to noise measurements of dense wavelength division multiplexing transmission systems, the readout of variety of optical sensors, and in the characterization of the transfer functions of passive and active photonic devices. Additionally, the spectral characteristics of different modulation formats can be analyzed, as well as the influence through nonlinear effects like FWM during the propagation through a fiber.

A simplified optical spectrum analyzer block diagram is shown in Fig. 4.1. The incoming light, or the signal under test (SUT), passes through a wavelength-tunable optical filter which resolves the individual spectral components. Afterwards, a photo diode (PD) converts the optical signal to an electrical current proportional to the incident optical power. The current from the photo diode is converted to a voltage by the transimpedance amplifier (TIA) and

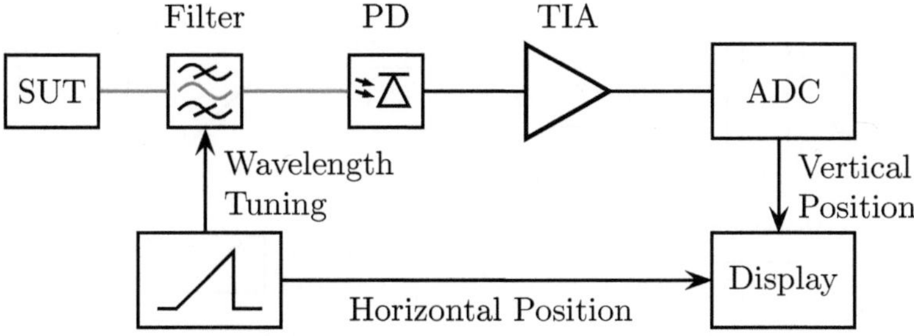

Figure 4.1.: Simplified block diagram of an optical spectrum analyzer.

subsequently digitized. Any remaining signal processing, such as applying correction factors, is performed digitally. The signal is then applied to the display as the vertical, or power data. A ramp generator determines the horizontal location of the trace as it sweeps from left to right. The ramp also tunes the optical filter so that its resonant wavelength is proportional to the horizontal position. This results in a trace of optical power versus wavelength. The ability of an optical spectrum analyzer to display two signals closely spaced in wavelength as two distinct responses is determined by the wavelength resolution. The minimum resolution is, in turn, determined by the bandwidth of the tunable optical filter, which can be realized by various mechanisms. Classically, optical spectrum analyzers can be divided into three categories: diffraction-grating-based and two interferometer-based architectures, which are the Fabry-Perot and Michelson interferometer-based optical spectrum analyzers [109].

The Fabry-Perot interferometer, shown in Fig. 4.2(a), consists of two highly reflective, parallel mirrors that act as a resonant cavity which filters the incoming light. The resolution and frequency range of Fabry-Perot interferometer based optical spectrum analyzers depends on the reflection coefficient of the mirrors and their spacing. Typically, these parameters are fixed and the wavelength sweep is done by changing the spacing between the mirrors by a very small amount. The major disadvantage is that the Fabry-Perot cavity lead to several pass bands depending on the free spectral range. Therefore, multiple wavelengths will be passed by the filter. However, this problem can be solved by placing an additional optical filter in cascade with the Fabry-Perot interferometer to suppress all power outside the interferometers free spectral range and achieve just the wavelength of interest. The advantage of the Fabry-Perot interferometer is its narrow spectral resolution. State of the art devices are capable of resolution in the range of 100 MHz. This narrow resolution allows them to be used for measuring laser chirp. But, their measurement span is much more limited than it is

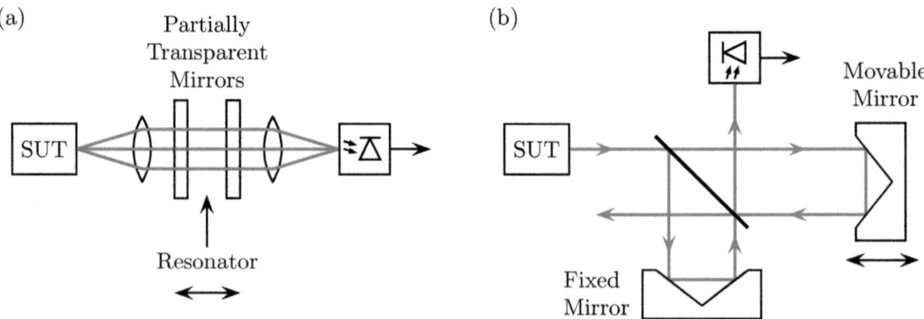

Figure 4.2.: Two typical interferometric structures for optical spectrum analyzer are the Fabry-Perot interferometer (a) and the Michelson interferometer (b).

the case for other types like diffraction-grating-based optical spectrum analyzers.

The Michelson interferometer, shown in Fig. 4.2(b), is based on creating an interference pattern between the signal and a delayed version of itself. The power of this interference pattern is measured for a range of delay values. The resulting waveform is the autocorrelation function of the input signal. This enables the Michelson interferometer-based spectrum analyzer to make direct measurements of coherence length, as well as very accurate wavelength measurements. Other types of optical spectrum analyzers cannot make direct coherence-length measurements. To determine the power spectra of the input signal, a Fourier transform is performed on the autocorrelation waveform. Because no real filtering occurs, Michelson interferometer-based optical spectrum analyzers cannot be put in a span of zero nanometers, which would be useful for viewing the power at a given wavelength as a function of time. This type of analyzer also tends to have less dynamic range than diffraction-grating-based optical spectrum analyzers. Nowadays, both interferometer based optical spectrum analyzers are hardly used and are not relevant in the field of optical communications.

The most common optical spectrum analyzers utilize a diffraction grating and a moveable aperture as tunable optical filter. A diffraction grating is a mirror with grooves on its surface, as shown in Fig. 4.3. The spacing between grooves is extremely narrow, approximately equal to the wavelengths of interest. When a parallel light beam strikes the diffraction grating, the light is reflected in a number of directions calculated by $\sin \alpha + \sin \beta = m\lambda/g$, where α and β are the input and output angle, m the order of the reflected beam and g the spacing of the grooves on the grating. If the input light contains more than one wavelength component, the beam will have some angular dispersion. Therefore, the reflection angle for each wavelength must be different in order to satisfy the requirement that the path-length difference off adjacent grooves is equal to one wavelength. Thus, the optical spectrum analyzer separates

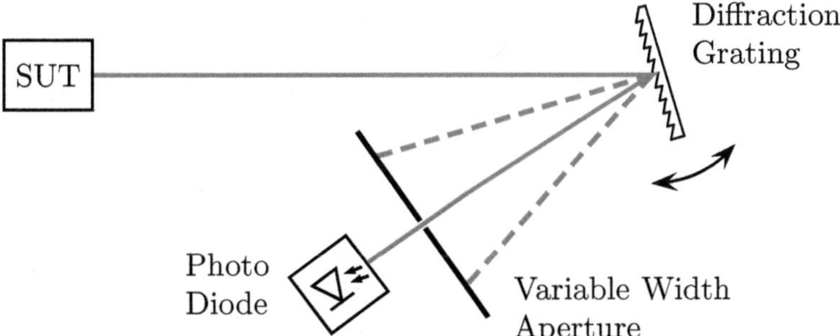

Figure 4.3.: Diffraction-grating based optical spectrum analyzer.

different wavelengths of light. The diffracted light passes through an aperture to the photo diode, as illustrated in Fig. 4.3. As the diffraction grating rotates, the instrument sweeps a range of wavelengths, allowing the diffracted light, the particular wavelength depends on the position of the diffraction grating, to pass through to the aperture. This technique allows the coverage of a wide wavelength range. Diffraction-grating-based optical spectrum analyzers contain either a single monochromator, a double monochromator, or a double-pass monochromator. One monochromator consists of an concave mirror at the input which parallelizes the light in order to use the full area of the grating that separates the different wavelengths and a second concave mirror which focuses the desired wavelength of light at the aperture. The aperture width is variable and is used to determine the wavelength resolution of the instrument. Diffraction-grating-based optical spectrum analyzers are widely used and capable of measuring all previously mentioned applications. The resolution of state of the art instruments is variable, typically ranging from 0.02 nm to 1 nm. This equals a minimum resolution of approximately 2.5 GHz at a wavelength of 1550 nm. All spectral components beneath remain hidden. The increased channel densities in dense wavelength division multiplexing optical communications systems, narrower laser sources as well as higher order modulation formats require spectral measurements with a resolution beyond that offered by conventional grating-based optical spectrum analyzers.

One powerful technique for obtaining higher spectral resolution is optical heterodyning, which has been in use since the invention of the laser. For many years, however, the use of optical heterodyne methods has been limited because of the narrow continuous-tuning range of lasers that can serve as local oscillators. With increasing availability of narrow linewidth and stable tunable laser source (TLS), an arbitrarily high spectral resolution may be obtained, at least in principle, through heterodyne detection. Therefore the signal under test is interfered with a local oscillator (LO), as can be seen in Fig. 4.4. The optical signal is detected with a photo diode and subsequently analyzed with an electrical spectrum analyzer (ESA). The signal and the local oscillator are superimposed at a the photo diode, which has

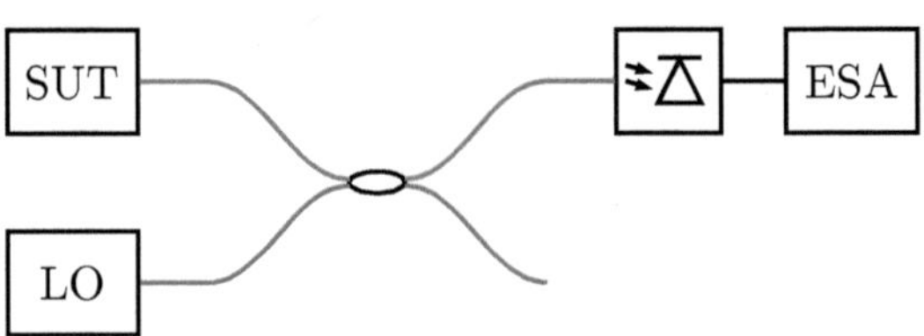

Figure 4.4.: Heterodyne detection, where the SUT is superimposed with an optical local oscillator.

a non-linear response to the amplitude. This means that at least a part of the output is proportional to the square of the input. The signal under test can be represented as [110]:

$$E_{Sig} = \cos\left(\omega_{Sig} t + \varphi\right) \tag{4.1}$$

and the local oscillator can be defined as:

$$E_{LO} = \cos\left(\omega_{LO} t + \varphi\right) \tag{4.2}$$

In general, the output current I of the detector is proportional to the square of the amplitude [2]:

$$
\begin{aligned}
I \propto\ & \left(E_{Sig} \cos\left(\omega_{Sig} + \varphi\right) + E_{LO} \cos\left(\omega_{LO} t\right)\right)^2 \\[2mm]
=\ & \frac{1}{2} E_{Sig}^2 \cos\left(1 + \cos\left(2\omega_{Sig} t + 2\varphi\right)\right) \\[2mm]
& + \frac{1}{2} E_{LO}^2 \cos\left(1 + \cos\left(2\omega_{LO} t\right)\right) \\[2mm]
& + E_{Sig} E_{LO} \left[\cos\left(\left(\omega_{Sig} + \omega_{LO}\right) t + \varphi\right) + \cos\left(\left(\omega_{Sig} - \omega_{LO}\right) t + \varphi\right)\right]
\end{aligned}
\tag{4.3}
$$

The output consists of different frequency ($2\omega_{Sig}$, $2\omega_{LO}$ and $\omega_{Sig} + \omega_{LO}$) and constant components. Since the current of the photo diode can not follow the high frequency components, the output reveals the beat frequency or intermediate frequency (IF) at $\omega_{Sig} - \omega_{LO}$. The amplitude of this last component is proportional to the amplitude of the signal radiation. The measured spectrum represents a convolution of the local oscillator with the signal. Therefore, the line width of the LO directly defines the resolution.

The signal spectra and the local oscillator in the optical domain can be seen in Fig. 4.5(a) and the detected signal in the electrical domain at the ESA is shown in Fig 4.5(b). As can be seen there is a signal part at lower frequencies, named direct detection. This part arises from the mixing of the signal itself. The higher frequency part is the heterodyned signal.

As can be seen, with increasing spectral width of the signal under test both parts will overlap at a certain point and the measurement will be distorted. Therefore, the maximum spectral width that can be detected at once is limited. Additionally, the achievable measurement range depends on the bandwidth of the photo diode as well as the bandwidth of the ESA. The method of heterodyne detection can be enhanced by the utilization of balanced detection [111, 112]. Therefore, the direct detection parts vanish and broader spectral ranges can be measured at once. By sweeping the LO even a broad spectral range can be measured, one part

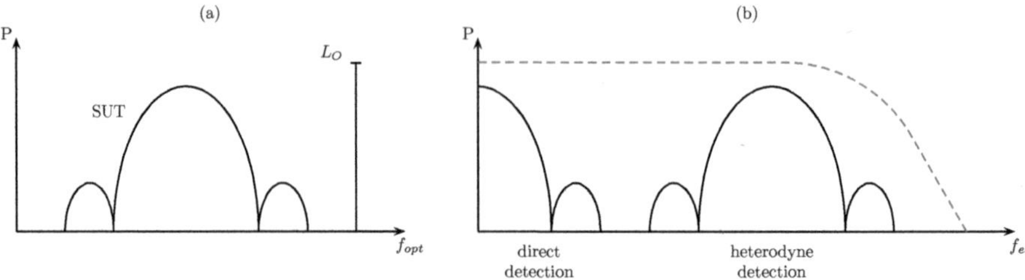

Figure 4.5.: Optical signals (a) and electrical signals (b) during heterodyne detection.

after the other. However, the practical realization of such coherent OSAs is often challenging, as they may require highly stable and low-noise local oscillators, optical phase-locked loops or frequency comb sources [111, 112], and so on. Measurement techniques which employ direct detection are in general simpler to implement. Additionally, the usable spectral measurement range of coherent OSAs is sometimes restricted by the bandwidth of the subsequent RF spectrum analyzer and that of the photo diode, which are on the order of several tens of GHz, as illustrated by the dashed gray line in Fig. 4.5(b). For ultra-wideband optical systems this measurement range might not be enough.

Over the last decade, high-resolution optical spectrum analyzers based on stimulated Brillouin scattering amplification were proposed and demonstrated [113, 114]. The Brillouin based optical spectrum analyzer (BOSA) represents a great step in the optical spectrum analysis technology, with an optical resolution of nearly three orders of magnitude better than conventional grating based optical spectrum analyzers. When the signal under test is measured with this powerful tool, the optical spectrum reveals details that have never been seen before.

4.1. Operation Principle

Compared to a classical OSA, the tunable optical filter is realized by the narrow band amplification of stimulated Brillouin scattering. Within a BOSA the pump wave and the SUT counter propagate along an optical fiber. The pump wave introduces a relatively narrow band spectral amplification of the signal wave, whose center frequency is down shifted with respect to the pump wave by the Brillouin shift of the fiber f_B, as can be seen in Fig. 4.6. When the pump is sufficiently strong, the overall signal power at the fiber output is dominated by the contribution of amplified spectral components that fall within the SBS gain window. Hence the signal power spectral density can be reconstructed through scanning the frequency of

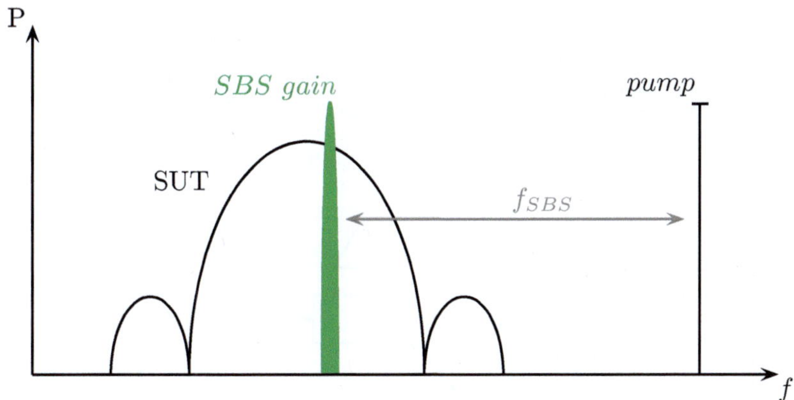

Figure 4.6.: Illustration of the frequency domain signals of a Brillouin based optical spectrum analyzer.

the pump wave [113, 114]. The fundamental resolution of this technique is restricted by the Brillouin bandwidth $\Delta\nu_B$, which can be in the order of 10 MHz in standard fibers at 1550 nm wavelengths, as shown in section 3.1.

The principle setup for a BOSA can be seen in Fig. 4.7. It consists of a laser source that can scan through the wavelength range of interest. The frequency sweep is realized by a coarse tuning of the pump laser itself and an additional external modulator that allows a very precise tuning by shifting slightly the sideband that generates the SBS gain. The pump wave is coupled into the fiber via a circulator. The signal under test is injected in the optical fiber from the opposite direction through an optical isolator. This will prevent any perturbation on the signal source due to the transmitted power from the pump wave that reaches the end of the fiber. This amplified signal is detected with a photo diode and stored through a conventional data acquisition setup. The photo diode basically acts as power detector and need no high bandwidth. During the measurement the power value for every frequency step is measured and displayed. The nature of the Brillouin interaction between pump and signal wave implies the need for a proper alignment of polarization states between both waves in order to get greatest efficiency in the amplification of the measured signal. This alignment is obtained by inserting polarization controllers before the fiber for the pump and the signal wave.

The comparison of the measurements between a classical grating based OSA and a BOSA can be seen in Fig. 4.8. Therefore the signal under test was a 2^{15}-1 pseudo random bit sequence (PRBS) with a data rate of 0.5 Gbps, modulated in a non return to zero format.

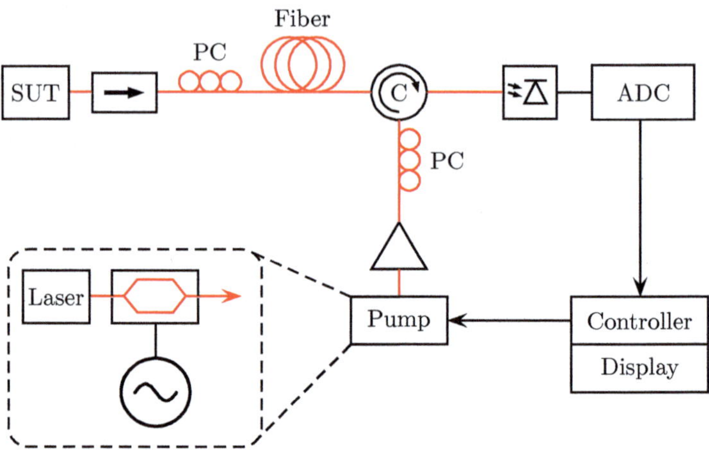

Figure 4.7.: Operating principle of a Brillouin based optical spectrum analyzer.

The measurement with the OSA is displayed in Fig. 4.8(a). As can be seen, with the grating based OSA neither the real shape of the spectrum nor any detailed information can be carried out, due to the limited filter bandwidth. The measurement with the BOSA can be seen in Fig. 4.8(b) and corresponds to the gray area in the OSA measurement [115]. The shape of the spectrum can be clearly distinguished. The advantage of the small filter bandwidth of the BOSA and thus the higher resolution is conspicuous.

One advantage of the BOSA is that the spectral analysis is done purely in the optical domain. Therefore, it prevents any overlapping from the modulation spectra or unwanted mixing products between optical sources, which is a serious drawback of heterodyning methods. Compared to the diffraction grating techniques used in a conventional OSA the finesse of the optical filter provided by SBS is significantly higher and can not be achieved by gratings. In addition, the amplification of the signal under test has a decisive influence on the simplicity of the electrical detection stage. But the BOSA is prone, however, to crosstalk from out-of-band spectral contents. Signal components outside the SBS gain window, although not amplified, propagate to the output with little loss, and thereby deteriorate the close-in dynamic range or optical rejection ratio, which signifies the ability to recognize a weak spectral component in the presence of a second, stronger one at a closely spaced frequency. This can be overcome by the utilization of the superposition of the gain with two losses and the simultaneous utilization of the polarization pulling effect of SBS, as shown in the forthcoming section.

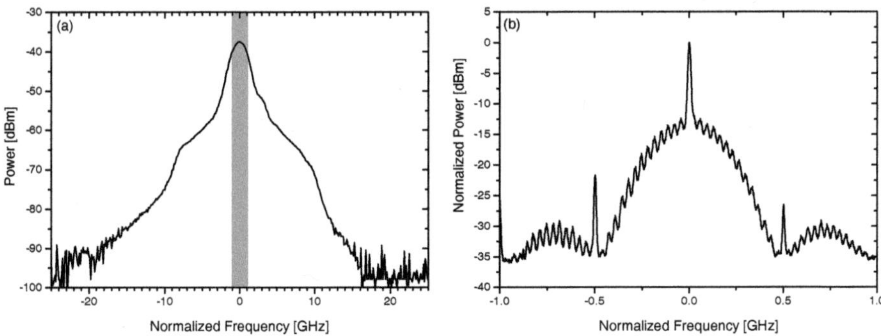

Figure 4.8.: Measurement of a 0.5 Gbps NRZ PRBS 2^{15}-1 with a conventional grating based OSA (a) and a BOSA (b).

4.2. Resolution and Dynamic Range Enhancement

As have been shown, the BOSA can achieve a resolution bandwidth of 10 MHz. In recent years, such resolution is increasingly becoming insufficient. For example, spectral efficient, modern optical communication formats, such as optical orthogonal frequency domain multiplexing (O-OFDM), make use of a large number of sub-carrier tones that are densely packed [116, 117]. The frequency separation between adjacent sub-carriers may be as small as a few tens of MHz [116], hence the monitoring of such signals would benefit from an OSA of comparable resolution or beyond. Similarly, a high resolution, wideband OSA would be instrumental in the characterization of dense radio-over-fiber transmission links [118, 119] and for millimeter-wave or terahertz communication systems [120]. Last but not least, a high resolution OSA would also help identify miniscule variations in the transfer function of resonant photonic devices, and thereby facilitate extremely sensitive environmental monitoring [121, 122]. For the measurement of laser linewidths, the signal under test can be used as a pump source for an ultra-narrow Brillouin fiber laser [123]. Heterodyne beating between the Brillouin fiber laser and the original signal yielded spectral measurements with a frequency resolution in the kHz range [123]. The spectral range of the measurement, however, was restricted to approximately 10 MHz in SSMF.

One possibility for the enhancement of the BOSA resolution is the utilization of dynamic Brillouin gratings. Thereby, an ultra-narrow bandwidth optical filter is realized by creating a long grating in the optical fiber due to the refractive index changes of the fiber induced by SBS [124]. As short description can be found in chapter 5. First spectral measurements with this technique have shown a resolution of 0.5 MHz (4 fm) [124]. However, they are very sensitive to any birefringence non uniformity in the fiber.

The second approach might be the utilization of a multistage system. As shown in section 3.2 the Brillouin bandwidth can be reduced down to 5.8 MHz in such a system. With additional stages the bandwidth can be reduced even further. However, all stages need to have the same Brillouin shift and every stage forms an individual amplifier that adds noise to the system. Therefore the realization of such a system is not further investigated in this work.

The third approach for enhancing both, the resolution and the optical rejection ratio of a Brillouin-based optical spectrum analyzer, is based on the combination of two techniques [125]. First, two Brillouin loss lines are superimposed upon a central Brillouin gain to reduce its bandwidth, as described in section 3.4. Already by just employing this method the resolution can be enhanced significantly [126, 127]. Second, the vector attributes of stimulated Brillouin scattering amplification in standard, weakly birefringent fibers are used to change the signal state of polarization, as described in section 2.4. The basic operation principle, is of course similar to a classical BOSA and can be seen in Fig. 4.9.

The signal under test is launched from the left side through an isolator and a polarization controller into the fiber, which acts as Brillouin medium. From the right side, three pump waves, one for the gain and two for the losses, are coupled into the system via a circulator. The relative frequencies with respect to the signal under test can be seen in the inset of

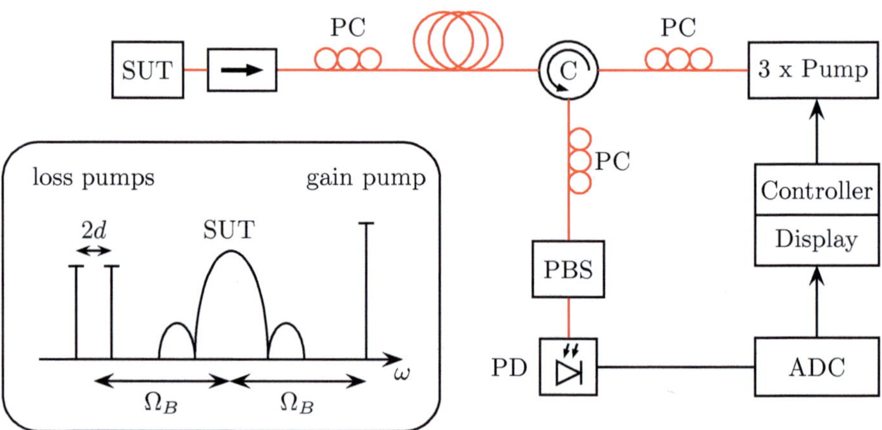

Figure 4.9.: Schematic illustration of a BOSA with enhanced resolution and dynamic range. The inset shows the relative frequencies of the pump waves with respect to the signal under test.

Fig. 4.9. This configuration would already lead to a significant reduction of the Brillouin gain bandwidth and therefore to a higher resolution of the BOSA. In accordance with section 3.4 the SBS gain coefficient of the spectral narrowing by the superposition of the gain with two losses can be rewritten as [125]:

$$
\begin{aligned}
g\left(\omega_{sig}\right) = {} & g_P P \left(1 - 2j\frac{\omega_{sig} - \omega_{sig}^0}{\Gamma_B}\right)^{-1} \\
& - m g_P P \left(1 - 2j\frac{\omega_{sig} - \omega_{sig}^0 - d}{\Gamma_B}\right)^{-1} \\
& - m g_P P \left(1 - 2j\frac{\omega_{sig} - \omega_{sig}^0 + d}{\Gamma_B}\right)^{-1}
\end{aligned}
\tag{4.4}
$$

where P denotes the power of the amplifying pump component, mP is the power of each of the two attenuating pump components, $\omega_{sig} - \omega_{sig}^0$ are optical frequency variables and g_P is the peak gain coefficient of SBS. The two detuned loss lines largely cancel out the gain to the sides of the primary amplification peak, rendering the linewidth of the entire composite process narrower [107].

The vector analysis of SBS had revealed that the SOP of the amplified signal is drawn towards a particular state, which is governed by the state of polarization of the pump, as described in section 2.4. Therefore, the SOP of the amplified spectral components of the signal under test differ substantially from the unamplified out of band components. It is therefore possible to discriminate further between amplified and unamplified spectral components of a broadband signal by utilizing a properly aligned polarizer. This method can significantly enhance the dynamic range of the BOSA. Based on this principle, a polarization beam splitter (PBS) is used to reject the unamplified spectral components, while those within the SBS bandwidth are partially transmitted, as can be seen in Fig.4.9. Therefore the polarization is aligned carefully with help of a polarization controller (PC). Finally, the amplified signal is detected with a narrow band photo diode, which acts in principle as a power detector. During the measurement of the spectra, the wavelength of the pump laser diodes is shifted by a controller and the according power value for each wavelength is measured and displayed.

The transmission axis of the polarization beam splitter at the fiber end is aligned with a state of polarization whose unit Jones vector is $\hat{p} = p_{max}\hat{e}_{max}^{out} + p_{min}\hat{e}_{min}^{out}$, with $|p_{max}|^2 + |p_{min}|^2 = 1$ and in accordance with section 2.4. The spectral distribution of the amplified signal at the output of the polarization beam splitter can be expressed by [47, 125]:

$$
\left|\vec{E}^{out}\left(\omega_{sig}\right)\right|^2 = \left|E_0\left(\omega_{sig}\right)\right|^2 \left|a p_{max}^* G_{max}\left(\omega_{sig}\right) + b p_{min}^* G_{min}\left(\omega_{sig}\right)\right|^2
\tag{4.5}
$$

For any configuration of input SOPs of the signal and pump wave the output polarizer needs to be adjusted in a way that all out of band spectral components are rejected entirely, which means $ap^*_{max} + bp^*_{min} = 0$. Subject to this adjustment, the output signal power spectral density becomes:

$$\left| \vec{E}^{out} \left(\omega_{sig} \right) \right|^2 = \left| ap^*_{max} \right|^2 \left| G_{max} \left(\omega_{sig} \right) - G_{min} \left(\omega_{sig} \right) \right|^2 \left| E_0 \left(\omega_{sig} \right) \right|^2$$

$$\equiv \left| H \left(\omega_{sig} \right) \right|^2 \left| E_0 \left(\omega_{sig} \right) \right|^2$$

(4.6)

As can be seen, the transfer function for the amplified signal at the output of the polarization beam splitter is given by $H \left(\omega_{sig} \right) = ap^*_{max} \left[G_{max} \left(\omega_{sig} \right) - G_{min} \left(\omega_{sig} \right) \right]$. Although Brillouin amplification is a nonlinear process, it can be considered as linear in the signal wave within the undepleted pump regime [4]. Therefore, the adopted description in terms of an equivalent frequency domain transfer function is legitimate. The maximum value of the coefficient $\left| ap^*_{max} \right|$ is 0.25, with $\left| a \right|^2 = \left| p^*_{max} \right|^2 = \frac{1}{2}$, assuming the complete out of band rejection. For a sufficient strong pump wave, and in the vicinity of the gain peak it can be approximated [47, 125]:

$$\left| H \left(\omega_{sig} \right) \right|^2 \approx 0.25 \left| G_{max} \left(\omega_{sig} \right) \right|^2 = 0.25 \exp \left\{ \frac{2}{3} Re \left[g \left(\omega_{sig} \right) \right] L \right\}, \qquad (4.7)$$

whereas outside the Brillouin amplification it yields $\left| H \left(\omega_{sig} \right) \right|^2 \approx 0$. This behavior is in significant contrast to the transmission of unamplified spectral components in a scalar Brillouin amplification process, without polarization assistance, where $\left| H \left(\omega_{sig} \right) \right|^2 \approx 1$. Therefore, the polarization pulling enhanced BOSA will provide a considerably larger optical rejection ratio than that of a corresponding scalar arrangement. In addition, the polarization discrimination assists reducing the spectral width of $H \left(\omega_{sig} \right)$ below that of a scalar Brillouin amplification window.

The calculated transfer function for Brillouin amplification of the signal under test, as a function of the frequency offset $\Delta f \equiv \left(\omega_{sig} - \omega^0_{sig} \right)$, can be seen in Fig. 4.10. The used parameters were $L =20 \, \text{km}$, $\Gamma_B =40 \, \text{MHz}$ and $g_0 = 0.1 [m \cdot W]^{-1}$. The three pump wave process was calculated by Eqn. 4.4 with the parameters $P =9 \, \text{mW}$, $m = 0.45$, and $d = 0.3$, as illustrated by the red lines in Fig 4.10. Accordingly, the single gain Brillouin amplification, represented by the black lines, was calculated with $m =0$ and a power of $P =4 \, \text{mW}$. The dashed lines represent the normal scalar Brillouin process, calculated by $\left| H \left(\omega_{sig} \right) \right|^2 = \exp \left\{ \frac{2}{3} Re \left[g \left(\omega_{sig} \right) \right] L \right\}$, and the solid lines show the polarization pulling assisted process calculated through Eqn. 4.6.

As can be seen in Fig. 4.10, the transfer function of the polarization assisted SBS processes effectively rejects the out-of-band components of the signal under test. In addition, the transfer function of the composite, three-pump process is seen to be narrower than that of a

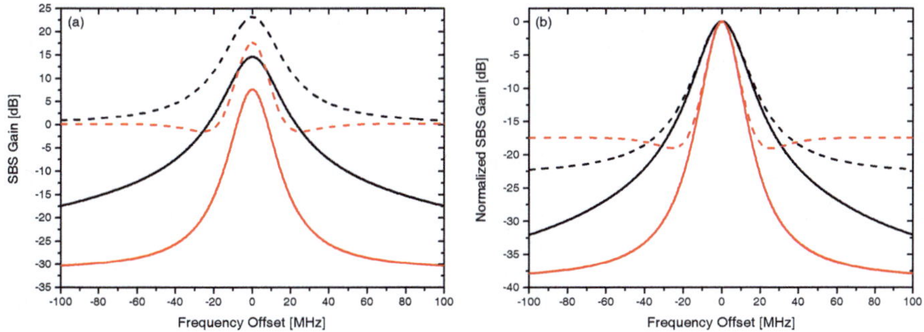

Figure 4.10.: Calculated transfer function $|H(\Delta f)|^2$ (a) and the normalized power gain $|H(\Delta f)|^2 / |H(\Delta f = 0)|^2$ (b) for various processes. The dashed lines show the scalar SBS process and the solid lines illustrate the polarization assisted process, whereas the the black lines show the one pump system and the red lines a three pump wave configuration.

single SBS gain line and it suppresses the out-of band components even further. Compared to the corresponding scalar process, the triple-pump, polarization-assisted Brillouin amplification provides relative out-of-band suppression that is 20 dB stronger.

The experimental setup is shown in Fig. 4.11. As in the simulation, a 20 km long AllWave fiber was used as the SBS gain medium. The SBS amplifying pump was driven by an ultra-narrow fiber laser (FL) of 1550 nm wavelength and a linewidth below 2 kHz with the optical frequency ω_p. An erbium-doped fiber amplifier (EDFA1) and a manual polarization controller (PC3) were used to adjust the power and SOP of the amplifying pump, respectively, as shown in the yellow frame in Fig. 4.11. The two pump waves for the Brillouin losses were generated independent by a DFB laser diode with a linewidth of 1 MHz. The central frequency was adjusted to $\omega_p - 2\Omega_B$ and stabilized through careful current and temperature control. The light from the DFB laser diode output passed through a Mach-Zehnder modulator (MZM4) which is driven by a sine-wave of frequency d that specifies the distance of the losses. The modulator operates in the suppressed carrier regime and the resulting sidebands at the frequencies $\omega_p - 2\Omega_B \pm d$, were amplified by EDFA2 and used as the Brillouin loss pumps. The state of polarization of the loss pumps is aligned in parallel with that of the gain pump with help of PC4, as illustrated in the blue frame of Fig. 4.11. The three pump waves were combined by a 3 dB coupler and launched into the optical fiber via a circulator (C).

During the proof of concept experiment the signal under test was generated also by the FL. In order to shift the signal under test into the gain region and for scanning the signal during the measurement it is processed by MZM1. This modulator was biased to suppress

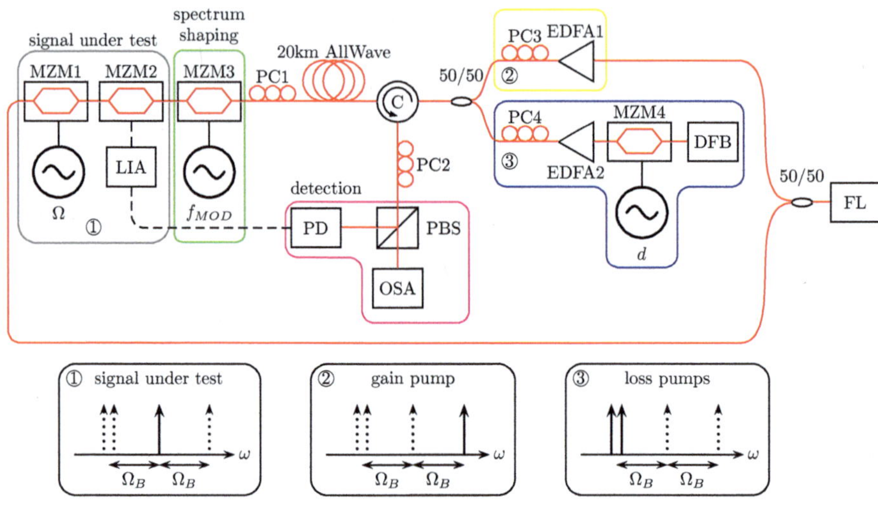

Figure 4.11.: Experimental setup for the realization of an ultra-high resolution optical spectrum analyzer based on the bandwidth reduction of the gain with two losses and additional polarization pulling. The insets (1) through (3) denote the relative frequencies of the signal under test, the amplifying Brillouin pump and the attenuating Brillouin pumps, respectively. The signal under test consists of the modulation sidebands as well as low frequency components from the lock-in amplifier.

the optical carrier at frequency ω_p, and was driven by a sine wave at varying frequency $\Omega \sim \Omega_B$ to generate double sidelobes of optical frequencies $\omega_p \pm \Omega$. The lower sideband was subject to potential amplification by the three-pump SBS process, depending on the choice of Ω. The relative frequencies of the pump wave and the signal under test can be seen in the insets of Fig. 4.11. The test signal was then modulated in two stages. First, low frequency modulation from the output of an electrical lock-in amplifier (LIA) was imposed on the signal through MZM2, in order to facilitate the low-noise detection of weak optical power levels which becomes necessary during the experiment. The generation of the signal under test is illustrated by the gray frame in Fig. 4.11. Next, MZM3 was used to introduce modulation sidebands, with variable magnitude and variable frequency spacing f_{mod} in the range of 3–30 MHz, as highlighted by the green frame in Fig. 4.11. These sidebands were used to assess the resolution and optical rejection ratio of the polarization-assisted BOSA. For the sake of simplicity the signal under test and therefore the spectrum is swept through the fixed narrowed gain through varying Ω. Although a practical BOSA would work in the opposite manner, with the signal held fixed and the frequencies of the pumps being swept, the chosen arrangement was more readily accessible within the equipment constraints, and is nevertheless equivalent in terms of dynamic range and resolution. Following the multiple modulation stages, the test signal was injected into the fiber counter propagating to the pumps. The state of polarization of the signal under test is adjusted at the fiber input with PC1. The output signal of the fiber is passed through PC2 and a PBS, operating together as an output polarizer of adjustable state. The output power of the signal under test was measured by a photo-diode PD, and connected to the input port of the lock-in amplifier. This detection is illustrated in the magenta frame in Fig. 4.11). A commercial OSA was connected to the complementary output of the PBS for monitoring the position of the lasers during experiment.

Special attention was given to the relative alignment of the multiple polarization controllers. The launch SOPs of all three pumps (PC3, PC4) were co-aligned along an arbitrary state, and held fixed for the entire duration of the experiment. For initial calibration, the input signal was disconnected, and the amplified spontaneous Brillouin scattering emission was observed at the signal output end. The SOP of the spontaneous Brillouin scattering is known to be aligned with $\hat{e}^{out}_{max}$, as discussed in section 2.4. First $\hat{e}^{out}_{max}$ was identified by setting PC2 to maximum transmission of the spontaneous Brillouin scattering through the PBS to the PD, and the output power was noted. Then, PC2 was readjusted until the transmitted power of the spontaneous Brillouin scattering was reduced by 50%, signifying $|p^*_{max}|^2 = 1/2$. Lastly, the test signal was reintroduced, and PC1 was used to adjust the test signal input SOP until the output signal was entirely blocked by the PBS in the absence of SBS amplification. Therefore, Ω was detuned temporarily by several times Γ_B. This final adjustment guaranteed that $ap^*_{max} + bp^*_{min} = 0$, as necessary.

During the measurements with the normal SBS gain the output pump power set at EDFA1 was 4 mW. For the experiments with three pump waves, the power of the gain pump was raised to 9 mW and the pump power of the losses was set to 4 mW. As shown previously, with this parameters and the given fiber length significant polarization pulling is expected. The frequency separation $2d$ between the two loss pumps was 16 MHz.

At the beginning of the measurement the spectral shape and the width of different configurations were characterized. During this type of measurement the modulation frequency f_{mod} at MZM3 was turned off. The obtained gain spectra for different configurations can be seen in Fig. 4.12. The measured full width at half maximum (FWHM) of the normal SBS process generated by a single pump wave was 40 MHz, as can be seen by the black curve in Fig. 4.12. The maximum gain of a single-pump, polarization assisted process was slightly lowered, however its bandwidth was reduced to around 20 MHz, as represented by the red curve in Fig. 4.12. As can be seen, the spectral components in the vicinity of the SBS gain

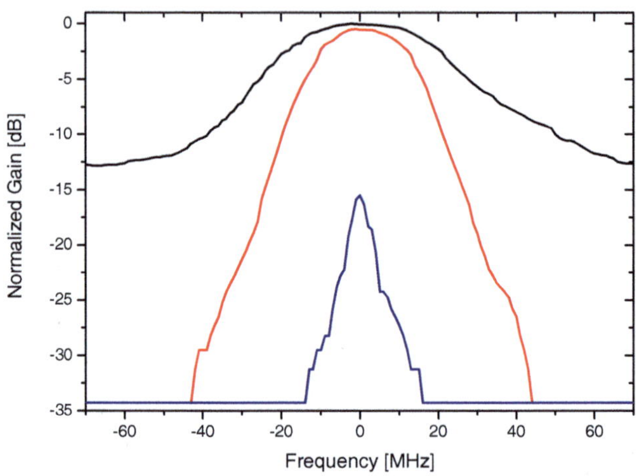

Figure 4.12.: Normalized measurement of the gain distribution for several SBS processes as a function of the detuning of the probe wave from its frequency of maximum amplification. The scalar process (black) attain a FWHM bandwidth of 40 MHz and the polarization assisted process (red) achieve 20 MHz. The three pump process (blue) reduce the bandwidth down to 3 MHz.

line were sharply cut off. The relative power gain at spectral detuning of 20 and 30 MHz was -10 dB and -20 dB, respectively. For detuning of more than 43 MHz from the line center, the spectral components of the signal were suppressed below the detection threshold of the lock-in amplifier. The introduction of the two Brillouin loss pumps, whose polarization is aligned parallel to the gain pump, lowered the gain for the composite SBS process by 15 dB. However, the gain bandwidth was reduced to around 3 MHz. The dynamic range of the measurements was restricted by that of the lock-in amplifier.

In order to test the spectral resolution and the dynamic range of the method, the frequency f_{mod} is turned on and forms the spectrum of the signal under test. The first measurements were carried out using polarization enhancement and a single gain pump wave. The results for several test signals, in comparison with a spectrum obtained by the process without polarization pulling, can be seen in Fig. 4.13. During these experiment MZM3 was driven by a sine wave with the frequency f_{mod}=30 MHz, which leads to sidebands at $\pm f_{mod}$ and $\pm 2f_{mod}$. Additionally, the voltage of the modulation signal was changed during the measurement in order to influence the modulation depth of the second sideband. As can be seen from Fig. 4.13, the normal SBS process could barely resolve even the first-order modulation sidebands at $\pm f_{mod}$. The optical rejection ratio of a spectral component detuned by 60 MHz was only

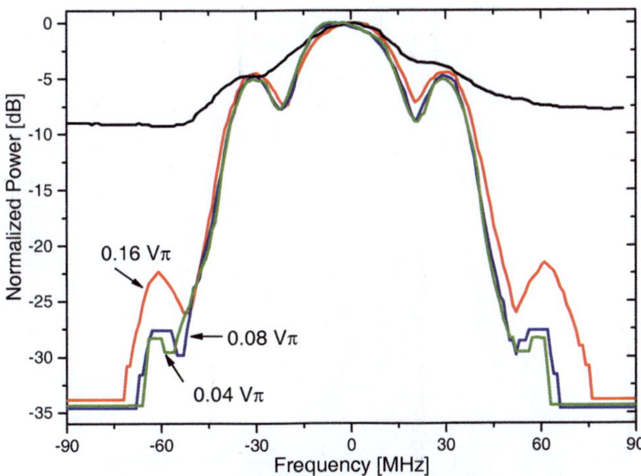

Figure 4.13.: Enhancement of the optical rejection ratio of BOSA by the polarization pulling effect. Blue, red and green curves correspond to polarization-enhanced measurements of signals that were modulated using different RF power levels. The black curve shows a measurement with the scalar SBS process.

9 dB, as illustrated by the black curve. With employed polarization enhancement, the first order sidebands at $\pm f_{mod}$ can be clearly distinguished from the carrier, and the second order sidebands were visible as well. Even for a very low modulation voltage of just $0.04\,\mathrm{V}_\pi$ the second order sidebands, which are 60 MHz away from the carrier and around 30 dB lower, could still be resolved.

The measurements of different spectra, obtained with polarization pulling in conjunction with bandwidth reduction through a three pump wave system are shown in Fig. 4.14. During this measurement f_{mod} was varied from 10 MHz down to 3 MHz. Even for a modulation frequency of as low as 3 MHz, the different sidebands in the test signal spectrum can be clearly distinguished. The resolution and optical rejection ratio of Brillouin-based spectral analysis could be clearly enhanced by applying the polarization pulling effect in combination with the gain narrowing through the superposition of the gain with two losses.

However, the spectral measurement range of this method would be restricted to the order of $2\Omega_B$. All signal components that are down shifted in frequency by Ω_B with respect to the loss pumps are amplified as well. Also the effect of polarization pulling takes place for these signals. In principle, a standard tunable optical filter can be added to suppress widely detuned spectral components. For a single gain pump wave the measurement range might be just limited by the polarization mode dispersion (PMD) in the fiber, since it prevents

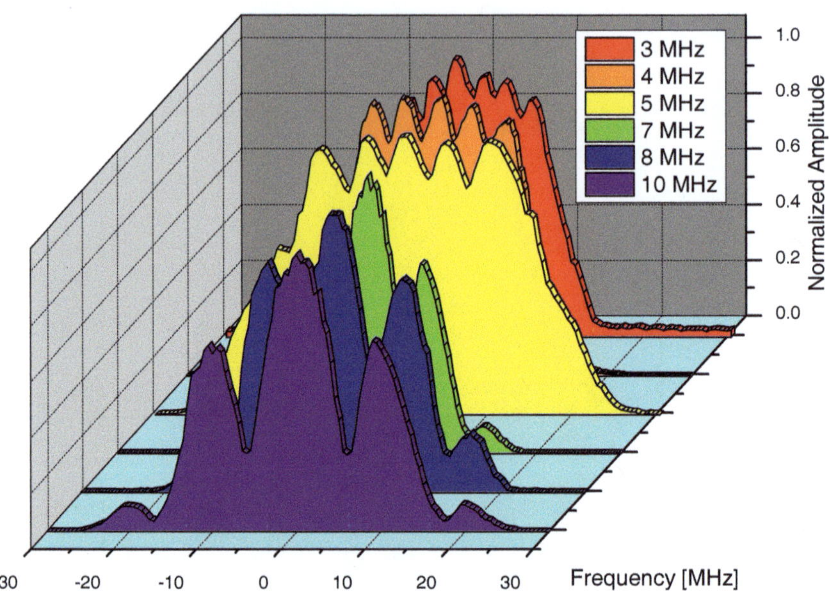

Figure 4.14.: Normalized, measured spectra of amplitude-modulated signals under test, on a linear scale. The modulation frequency was changed from 10 MHz down to 3 MHz.

the complete rejection of signal components that are widely detuned from the SBS gain peak. The differential group delay associated with PMD in standard fibers that are a few km long is on the order of picoseconds. Therefore, measurement ranges approaching the THz scale are feasible. The detrimental effect of PMD can be reduced if the signal under test is aligned with a principal state of polarization [128], and if the alignment of polarization component is re-calibrated at several frequencies. A broad spectral measurement range would also require an adjustment of the Brillouin frequency shift Ω_B as a function of the optical frequency of the signal under test [129]. Compared with an OSA based on a straight-forward, scalar SBS process, and, for the same low pump power levels (below 10 mW), the reported method provides an order-of-magnitude improvement in resolution to 3 MHz, and two-orders-of-magnitude enhancement of the optical rejection ratio to 30 dB. The dynamic range of GHz-resolution, grating-based OSAs, for example, could be as high as 80 dB. However, their optical rejection ratio for a detuning of several GHz is more modest, on the order of 30 dB. The relative SOPs remained sufficiently stable through a few minutes-long data acquisition. Polarization stability would benefit from the use of shorter fiber sections, and from more advanced electronic processing, leading to faster data acquisition.

4.3. Additional Utilization of a Local Oscillator

Besides the reduction of the SBS gain bandwidth, the resolution can be further enhanced by utilizing a local oscillator in combination with polarization pulling assisted stimulated Brillouin scattering (PPA-SBS) [130, 131]. This method for the ultra-high resolution analysis of optical signals is based on narrow band optical filtering in the first instance and a subsequent heterodyne detection with a narrow linewidth local oscillator. In principle, this technique combines the wide tuning range of Brillouin optical spectrum analysis and the suppression of the unwanted components through the polarization pulling effect with the high achievable resolution of heterodyne systems. In order to achieve a narrower filter for the preselection of the SUT, every mentioned method for the gain bandwidth reduction could be utilized. However, the following description will focus on the normal single gain.

The basic principle of the method is illustrated in Fig. 4.15. During the SBS process a pump wave generates a narrow gain for counter propagating signals which is down shifted in frequency by f_{SBS}, as described previously. All signal components inside the gain region are amplified. Thereby the SOP of the amplified signal is drawn towards the SOP of the pump wave, as described in section 2.4. Again the amplified components can be separated from the unamplified ones with a simple polarization filter like a polarization beam splitter or polarizer. This leads to just a small part of the spectrum under test which can now be

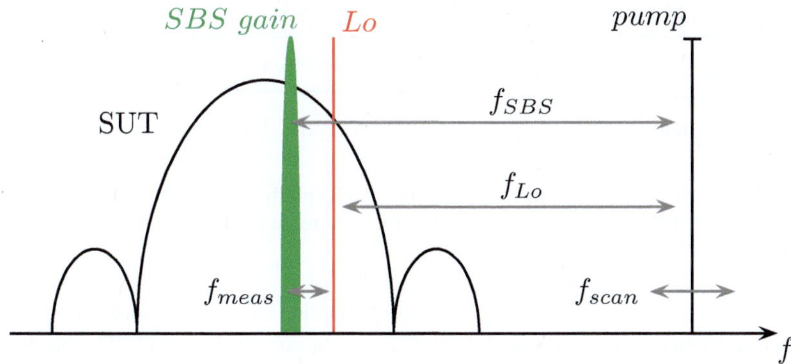

Figure 4.15.: Operation principle of the Brillouin optical spectrum analyzer assisted with heterodyne detection.

analyzed in detail with heterodyne detection. The LO is directly generated from the pump wave in order to simplify the setup and avoid drifting. In this regard, the pump wave is split and externally modulated with $f_{LO} < f_{SBS}$, which is a little smaller than the Brillouin frequency shift in the fiber. The down conversion to the electrical domain is done by a narrow bandwidth photo diode.

Thereby, unwanted mixing products from the signal components inside the gain region are down converted to the base band and therefore outside the measurement range of interest at the ESA, represented by the gray lines in Fig. 4.16. The wanted mixing products with the local oscillator arise around the frequency f_{meas}, depicted as red trace in Fig. 4.16. This frequency needs to be at least twice the Brillouin gain bandwidth in the fiber, in order to avoid overlapping of direct and heterodyne detection, and below the cut-off frequency of the photo diode. The electrical signal is finally analyzed with an ESA within a narrow span, shown as blue range in Fig. 4.16, at $f_{meas} = f_{SBS} - f_{LO}$. Therefore, just the heterodyned components in the center of the SBS gain are displayed and measured at the ESA. In order to measure the complete spectrum the pump wave, and respectively the gain and LO, is shifted through the unknown spectrum.

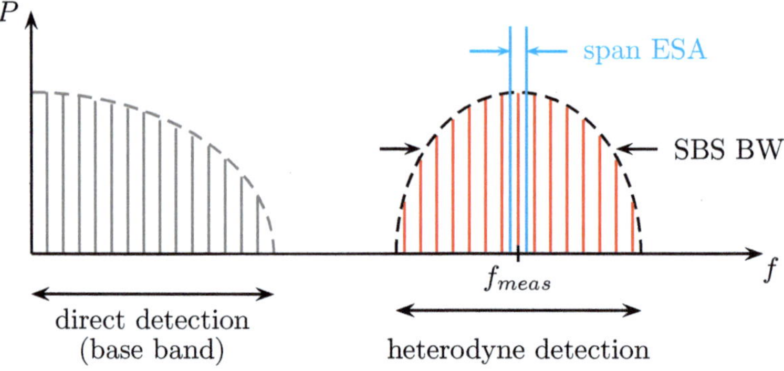

Figure 4.16.: Illustration of the mixing products at the photo diode (gray and red) after the preselection with PPA-SBS and subsequent heterodyning, and the final measurement range of the ESA (blue).

Exemplary, a measurement of the partial amplification of the SUT and suppression of the unwanted signal components with PPA-SBS can be seen in Fig. 4.17. The black curve represents the SBS gain. The red curve shows a small amplified spectral part of the SUT. Therefore, a 2^9-1 PRBS with a data rate of 1 Gbps was used. This results in a sinc shaped discrete spectrum with a frequency distance of $\Delta f = 1.95\,\text{MHz}$ between the lines. With

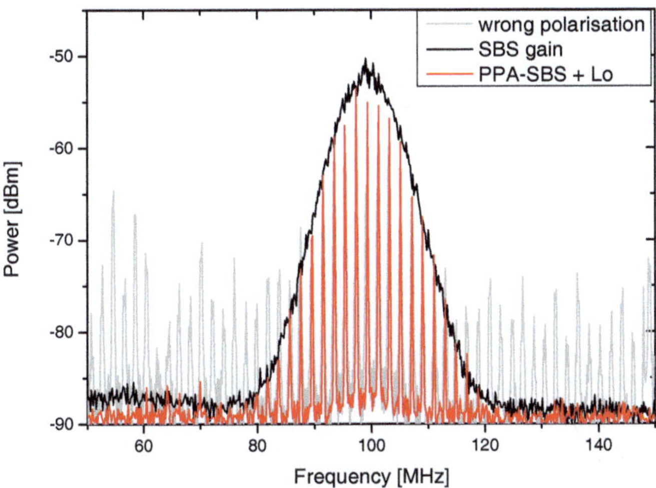

Figure 4.17.: The black curve shows the normal SBS gain in the fiber and the red curve implies the detailed measurement of a spectral part of the SUT from a 2^9-1 PRBS at a data rate of 1 Gbit/s. The wrong setting of the polarization, which leads to unwanted mixing products, is illustrated by the gray line.

a proper polarization alignment the unwanted out of band components can be suppressed, as illustrated in Fig. 4.17. The gray curve shows a misalignment of the polarization and therefore the result which would be achieved without polarization pulling. Due to the mixing in the photo diode this would lead to significant distortions.

The experimental setup for the first proof of concept experiments can be seen in Fig. 4.18. The main source for the system is a narrow linewidth fiber laser (FL). It acts as pump wave for SBS and local oscillator for the heterodyne detection. The PPA-SBS part of the setup is shown by the green box in Fig. 4.18. In order to shift the pump wave, as well as the LO, through the SUT, the laser is modulated externally with MZM1, which is driven in the carrier suppressed regime with f_{SBS}. In order to avoid unwanted interferences, the lower sideband is filtered out with a wave shaper (WS). Subsequently, the remaining sideband is amplified with an EDFA in order to provide sufficient power for the SBS. The polarization of the pump wave is aligned with PC2 in a way that it will pass through the PBS. The pump wave is split and one part is coupled into the fiber via a circulator (C).

In the lower path the SUT is injected into the system. During the experiments two different SUT were used. The first one was a femtosecond fiber laser, where the output spectrum is

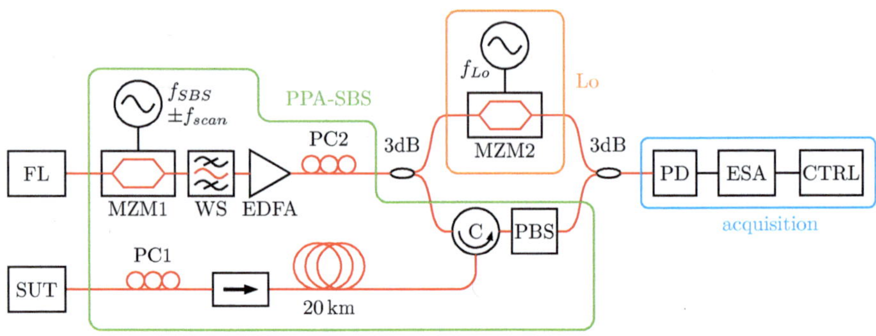

Figure 4.18.: Experimental setup for the Brillouin optical spectrum analyzer assisted by a local oscillator.

characterized. The second is directly generated out of the FL. Therefore, the signal of the FL is split at the beginning and processed by a phase modulator (PM) with different PRBS data patterns. The electrical power of the modulation signal is 11 dBm. The polarization of the SUT is aligned with PC1 in a way that it is blocked at the PBS. An isolator protects the laser source of the SUT. The SUT is coupled into the 20 km SSMF that acts as nonlinear medium for the SBS. From the other side the pump wave is coupled into the fiber. The partially amplified SUT is coupled out through the circulator to the PBS. Here the unwanted spectral components of the SUT are blocked entirely. The other part of the split pump wave is modulated with MZM2, driven by a sine wave (LO), which is shifted by 100 MHz with respect to f_{SBS} (orange box). This defines the final frequency f_{meas} where the signal is measured in the electrical domain. The LO is than combined with the preselected and amplified part of the SUT. The down conversion to the electrical domain is carried out with a PD with a bandwidth of 1000 MHz. Therefore, the mixing products at higher frequencies are cut off automatically. Finally, the electrical signal is analyzed with an ESA. The measurement takes place at $f_{meas} =$ 100 MHz with a span of 1 MHz. In this case just the center of the SBS gain distribution is measured. The unwanted mixing products in the base band are automatically rejected. In principle, the fixed span needs to be below the SBS bandwidth. For a zero span a simple power detector can be utilized. Alternatively, a narrow bandwidth electrical quartz filter could be used. It needs to have bandpass characteristics and a bandwidth in the range of the linewidth of the laser. For the measurement of the whole spectrum, the frequency of the sideband and consequently the LO is changed by the scan frequency f_{scan}. The scan process as well as the data recording is done by a separate controller (CTRL). The whole data acquisition is illustrated in the blue box in Fig. 4.18.

Within the first proof of concept experiments a 2^9-1 PRBS with a data rate of 500 Mbps was used, leading to a frequency distance of $\Delta f = 957\,\text{kHz}$ between the adjacent lines in the spectrum. For the measurement the SBS gain as well as the LO are shifted through the SUT by changing f_{scan} accordingly. In principle, the frequency is shifted by 10 MHz for every step and the data from the ESA is recorded. The span of the ESA needs to fit the step size of the frequency scan. Since the SBS bandwidth is larger than the actual span on the ESA, no distortions occur. The comparison of different measurement techniques can be seen in Fig. 4.19. The OSA can just detect the envelope of the SUT, depending on its resolution. The BOSA measurement was done in combination with a vector network analyzer [115] and reveals the sinc shaped envelope, but the distinct spectral lines the spectrum consists of, can not be displayed. With additional help of the LO the true characteristics of the SUT can be revealed, as can be seen in the black curve of Fig. 4.19.

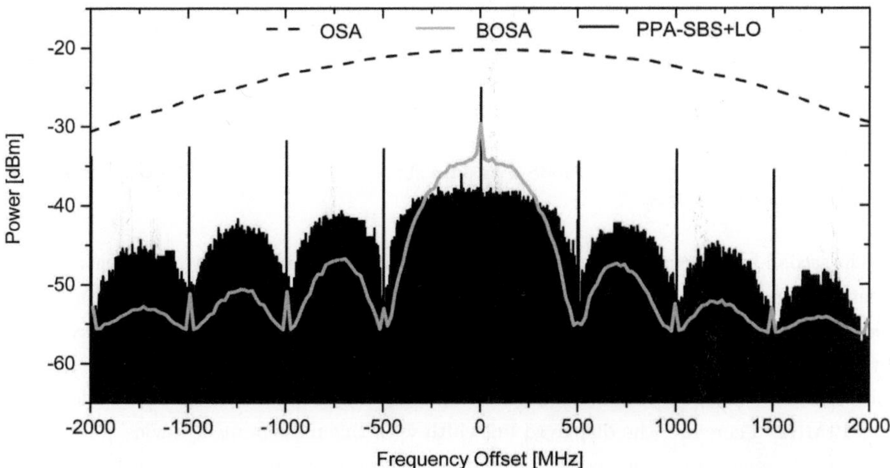

Figure 4.19.: Measurement results for PRBS 2^9-1 data signal at a data rate of 500 Mbps with different systems. The dashed curve shows the measurement with a conventional grating based OSA and the gray curve the measurement with a Brillouin OSA. The black curve shows the polarization pulling assisted process with additional local oscillator, where all details of the spectrum under test are revealed.

The complete spectrum for the proposed method is carried out by shifting the pump wave and respectively the local oscillator through the unknown spectrum. Therefore, the modulation frequency for the first modulator is adapted with an offset to f_{SBS} which covers the full range of the spectrum. However, for much broader spectral data the center wavelength of the laser diode needs to be shifted and stabilized. Additionally, narrow linewidth tunable lasers can be used. The minimal resolution is defined by the linewidth of the local oscillator. In the first proof of concept experiment the resolution was limited to 30 kHz. A detailed measurement of the spectrum under test is shown in Fig. 4.20. The different lines of the PRBS spectrum can be clearly distinguished.

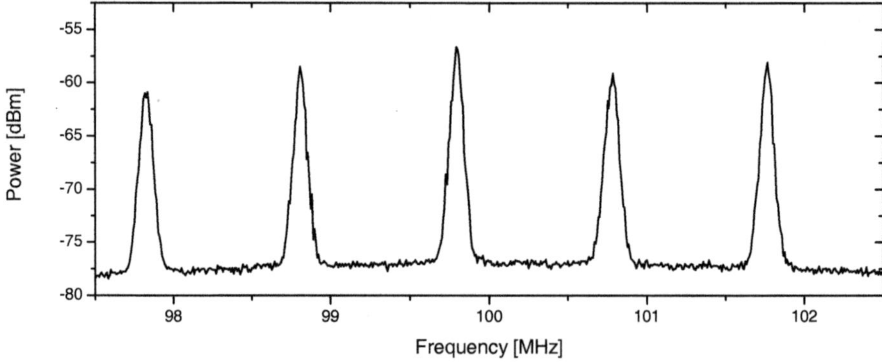

Figure 4.20.: Detailed part of a PRBS 2^9-1 data signal at a data rate of 500 Mbps measured with the enhanced BOSA.

The second measurement was carried out by investigating the frequency comb generated by a femtosecond fiber laser. The output spectrum measured with a conventional OSA can be seen in Fig. 4.21(a) and a detailed measurement was carried out with a BOSA again, as shown by the dashed curve in Fig. 4.21(b). The fs-laser has a repetition rate of 100 MHz. For the measurement with the BOSA the minimum resolution is restricted to the SBS bandwidth and was 10 MHz. Therefore, the displayed linewidth with this measurement device is incorrect. The measurement was repeated with the proposed method, and can be seen by the solid curve in Fig. 4.21(b). As local oscillator a fiber laser with a linewidth below 1 kHz was used. Through the relative slow measurement time with the high resolution, the laser drift can be clearly seen.

A detailed measurement of a single comb line at different wavelengths is shown in Fig. 4.22. As can be seen, the measured linewidth is approximately 6 kHz over the whole range. The fluctuations during the measurement might origin from the laser drift or power fluctuations. Since the linewidth of the LO is still much narrower than the measured line, it can be assumed that the measurement is correct. In principle, this figure shows the resolution limitations for the BOSA and the achieved resolution of the this method.

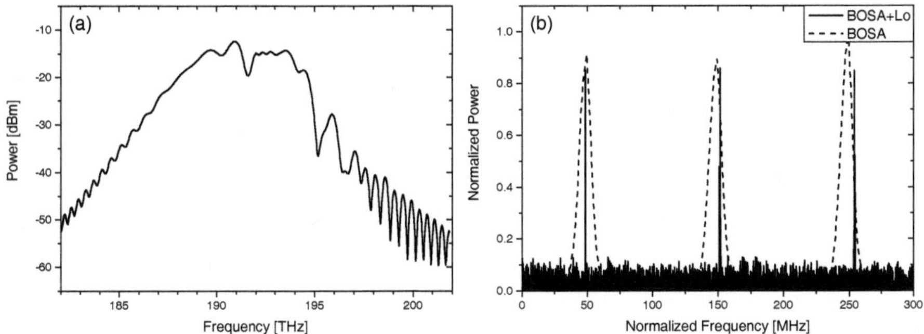

Figure 4.21.: Measurement of the power spectral density of a frequency comb generated by a fs-laser with a conventional OSA (a). Detailed measurements of several lines of the comb are provided by a BOSA (dashed line) and the introduced method (solid line).

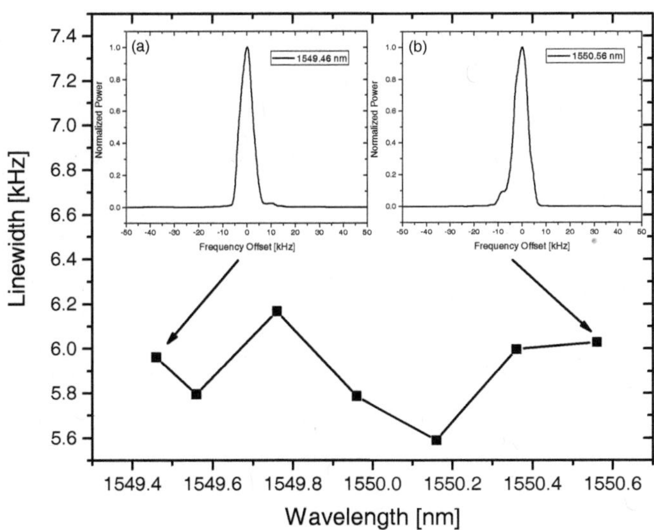

Figure 4.22.: Detailed measurement of the linewidth of a single mode of the fs-laser at different wavelengths.

Finally, it has been shown that the combination of narrow bandwidth filtering with stimulated Brillouin scattering with heterodyne detection can significantly increase the resolution of an optical spectrum analyzer. This enables the analysis of optical systems and components with ultra-high resolution over a spectral range which, in principle, is just restricted by the transparency range of the used Brillouin medium. The achieved resolution can be as low as $1\,\text{kHz}$ ($8\,\text{am}$ at $1550\,\text{nm}$), which is 2 million times better than a conventional grating based OSA. The method operates with off the shelf telecom equipment and enables the high resolution measurement of optical signals without any bandwidth limitation.

5

Delay and Storage of Light

Optical buffering is a major obstacle facing optical packet switching for future optical communication networking. The target application of optical memory is an all-optical router that temporarily stores data while identifying the address of a data packet, which is header recognition. Electronic buffers typically store megabits of data and the main argument in favor of optical schemes is that the required optical-electronic-optical conversion consumes a lot of power and limits the bandwidth of information. All-optical buffers operate, in principle, independently of the data rate and are more efficient, as the data remain in the optical domain. However, a bottleneck in modern optical communications is the lack of a fully tunable optical data buffer [132]. Therefore, the realization of all-optical buffers, without a conversion of the signals into the electrical domain, has been in the focus of several research activities during the last decades. This led literally to a race from the highest delay times, over the largest reduction of the speed of light down to 17 meters per second [133] and finally to the complete stopping and storage of light [134]. The outcome was a multitude of different approaches and methods for the realization of optical data buffers. A well known and increasingly evolved approach is a memory based on spatial spectral holography [135]. However, this method requires operation at cryogenic temperatures and is therefore impractical. Another approach for stored or stopped light is based on the electromagnetically induced transparency (EIT) [136], where the optical information is imprinted on the internal degrees of freedom of a dense atomic ensemble. These systems are characterized by impressive storage times of 1 s [137], but are experimentally very complicated, narrow band and the frequency of the stored light must be in the exact resonant frequency of the ions or atoms.

All the previously mentioned methods have the disadvantage that they can not be used or are difficult to use in optical communication networks or optical interconnects in the nanometer range. The utilization of the available propagation medium for the delay or storage of the optical data increases the chance of implementation into existing network structures

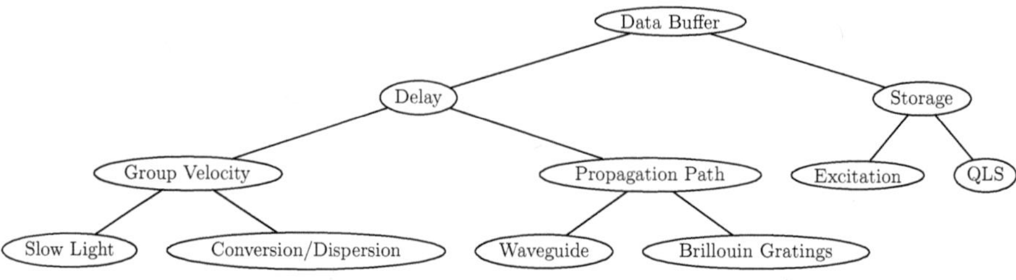

Figure 5.1.: Classification of several methods for the delay and storage of light in optical fibers.

economically. Therefore, in the following only methods that can be established within or with optical fibers are shown. In principle, they can be divided in two groups, as illustrated in Fig. 5.1. On the one hand there is the delay of the data signals and on the other hand the actual storage, where the signals are read into the memory and released again with the help of a read signal. The following sections will discuss both groups in detail and explain exemplary accompanying methods.

The category of delay in Fig. 5.1 can be divided further by means of its realization. On the one hand the group velocity can be manipulated and on the other hand the propagation path itself can be utilized for the delay of optical data.

It is commonly know that a monochromatic wave travels at the phase velocity given by [42]:

$$v_{ph} = \frac{\omega}{k(\omega)} \tag{5.1}$$

where $k(\omega)$ is the frequency dependent wave number and ω is an angular frequency. In a vacuum, the wave number is $k_0 = \omega/c$ and the phase velocity is independent of the wavelength and equal to the velocity of light. In a dispersive material the refractive index varies with the frequency. Hence, the phase velocity in the medium will be given by [42]:

$$v_{ph} = \frac{\omega}{k(\omega)} = \frac{c\omega}{n(\omega)\,\omega} = \frac{c}{n(\omega)} \tag{5.2}$$

An optical pulse, e.g. within data transmission, is generated by the interference of a large number of sinusoidal waves. In most materials the refractive index decreases with decreasing frequency, resulting in an increasing phase velocity. Hence, each of the waves has a different phase velocity. The velocity with which the pulse, in particular the pulse maximum, propagates is called **group velocity** and is given by [42]:

$$v_g = \frac{\partial \omega}{\partial k} \tag{5.3}$$

In general, the group velocity depends on the group index n_g, which is the refractive index for the group velocity. Combining Eqn. 5.2 and Eqn. 5.3 the group velocity of a pulse with the central frequency ω_0 can be written as:

$$v_g = \frac{c}{n\left(\omega_0\right) + \omega_0 \frac{\partial n(\omega)}{\partial \omega}} = \frac{c}{n_g} \tag{5.4}$$

In vacuum the group velocity equals the phase velocity and c because $n = 1$. However, in a dispersive medium like an optical fiber it will differ from v_{ph}. As can be seen from Eqn. 5.4, v_g does not only depend on the frequency-dependent refractive index but also on the frequency-dependent slope of the refractive index $\partial n\left(\omega\right)/\partial \omega$. If this slope varies with the frequency, the group velocity varies with the frequency as well. In a spectral region of normal dispersion where $\partial n\left(\omega\right)/\partial \omega > 0$, the group velocity decreases and the wave packet is delayed. If the refractive index slope is negative the group velocity is increased, leading to an acceleration of the wave packet. Therefore, the aim is to achieve the greatest possible variation of the group refractive index. Therefore a system is needed, which is highly dispersive over a narrow frequency range. However, this frequency range must by higher than the bandwidth of the signal; otherwise the wave packet will be distorted.

Starting a decade ago, there was a great research interest in the so called **Slow-Light** based on stimulated Brillouin scattering [138, 139]. Thereby, the dispersion in the fiber is artificially changed by the interaction between an incident pump wave and an acoustic wave in the medium. As shown in Chap. 2, the acoustic wave is generated by the pump wave itself and part of the pump power is scattered back. During the SBS process a gain is formed, where a counter propagating signal is amplified, as well as, a loss where the counter propagating signal is attenuated. The Brillouin gain and loss are associated with a phase change, as shown in Fig. 2.8, which in turn reflects in an artificial dispersion and thus in a change in the group refractive index n_g. For the gain this change is positive. Accordingly, the group velocity of the pulses is reduced and slow-light is formed. A disadvantage for the delay of higher data rates is the rather low SBS bandwidth of 10–30 MHz, depending on the used fiber and pump power [92], and therefore the area where Slow-Light can be achieved. Thus, the data rates in optical telecommunications for Slow-Light are greatly limited. In recent years, several methods for broadening the Brillouin bandwidth have been developed, such as the broadening of the laser linewidth of the pump laser [140] or the use of multiple closely spaced pump waves [141]. With these methods a Brillouin bandwidth of 12 GHz and the delay of 100 ps wide pulses by 3 bits have been reached [142]. However, the maximum achievable delay is limited [143]. Even by adjusting the shape of the Brillouin gain, only a maximum storage time of 4 Bit can be achieved [144]. But the almost distortion free delay of optical data signals has been reported [145].

Much higher delays may be achieved with the so called **Conversion/Dispersion** method [146, 147]. The basic principle is shown in Fig. 5.2. Thereby, the center wavelength of the input signal is shifted at the beginning via a non-linear optical conversion. Afterwards, the signal is passed into a highly dispersive fiber. The difference in group velocity in the dispersive fiber leads to a delay that is proportional to the product of the wavelength shift and the group velocity dispersion in the fiber. Subsequently, the original wavelength is restored by another conversion. For the conversion the nonlinear processes of self-phase modulation [148] or four-wave mixing [149] can be used. Previous experiments achieved a delay of 22 ns for a 2.6 ps wide pulse [150]. Additionally, delays of 243 ns for a 10 Gb/s NRZ data signal were obtained by phase conjugation, and the very large wavelength shift found in silicon waveguides [151]. This corresponds to a delay of 2400 Bit.

A disadvantage of this method is that for large delay times a significant dispersion and a large wavelength change is required, which can only be produced with considerable effort. Therefore, the Conversion/Dispersion method is one of the most complex and energy consuming methods for the delay of light and optical data [152].

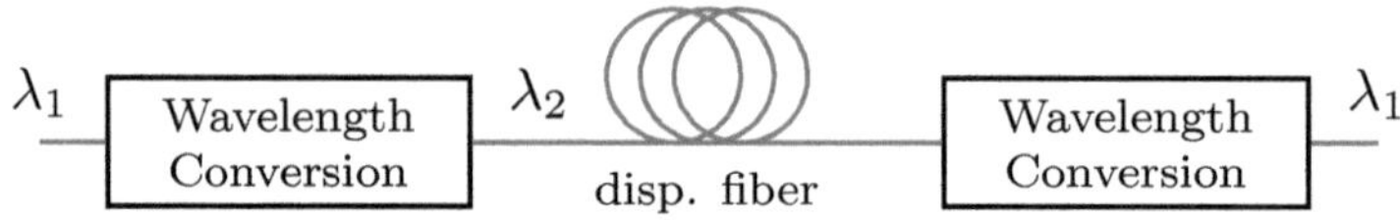

Figure 5.2.: Operation principle of the Conversion/Dispersion method.

As mentioned above, a delay is also possible via the **propagation path**. The easiest way to realize this are **waveguides** themselves [153], where the time shift of the signal is realized by its propagation time through the medium. The delay time is obtained from the quotient of length and the propagation velocity in the medium. Two possible implementations are illustrated in Fig. 5.3. In the simplest case, the signal is passed through a switch and delayed by an additional longer waveguide. By cascading several elements with waveguides of different lengths, various delay times can be achieved [154]. However, the delay times are not adjustable. They are fixed by design of the waveguide length. In general, the length of the delay line is adapted to the used packet length. The delay time can be increased by the utilization of a feedback and therefore the multiple use of the same element [155], as shown in Fig. 5.3(b). The signal runs through the delay path until it is switched to a different path. The different delay times and the maximum delay is determined by the number of rounds. However, no variable delays are possible here as well, since the delay time is just a multiple of a single element.

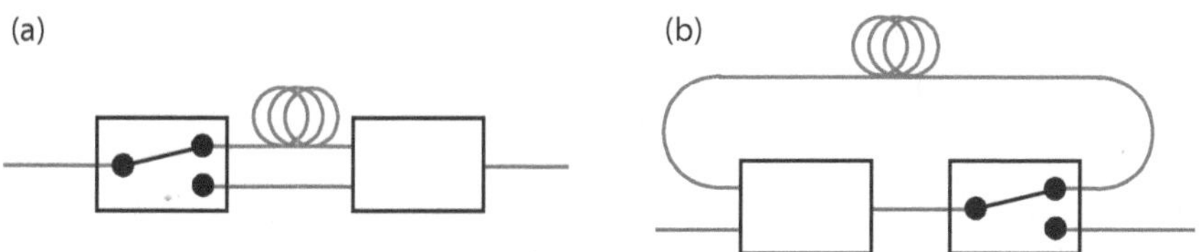

Figure 5.3.: Different arrangements of delay lines. (a) shows a standard system and (b) a system with feedback.

Another method, which utilizes the length of the propagation path for delaying optical signals, is based on **dynamic Brillouin gratings** [124]. If the SBS occurs in a birefringent medium such as a polarization maintaining fiber, the acoustic waves that are generated by optical waves from one polarization direction can be used as a reflection grating for frequency-shifted, orthogonally polarized optical waves. This reflection grating is referred to as dynamic Brillouin grating and emerges from the polarization-independent nature of the longitudinal acoustic waves [156]. The basic operation principle of a dynamic Brillouin grating (DBG) can be seen in Fig. 5.4. The two pump waves for the Brillouin scattering are counter propagating in x-polarized direction within a polarization maintaining fiber. The distance between both corresponds to the Brillouin shift within the fiber. If a signal propagates in y-polarized direction and in the same direction as pump wave 1, but frequency shifted by 50 GHz, it will be reflected at the acoustic waves. The frequency shift of the signal depends on the average refractive index and the average optical frequency of the pump waves [156].

The position of the grating in the fiber can be realized through a frequency shift of the pump waves. Thus the signal is reflected at different positions, and therefore the delay time is changed. Previous experiments were performed with a fiber length of 15 m and reached maximum delay times of 132 ns for Gaussian pulses with a pulse width of 82 ns [124]. This corresponds to a delay of 1.6 Bit. Theoretically, the storage time can be increased with the length of the fiber and remains variable. Though, for lower pulse widths significant distortions occur, which limits the utilization of this method in communication networks significantly.

The second path of the tree diagram in Fig. 5.1 deals with the storage of light. The storage

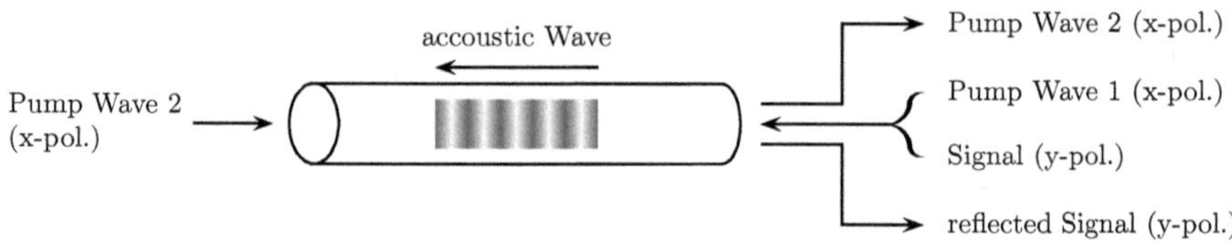

Figure 5.4.: Operation principle of a Brillouin grating in a polarization maintaining fiber [124].

of light in an optical fiber is possible by durable acoustic **Excitations** through the process of SBS, in which the optical pulse is converted into a slow moving acoustic wave and is converted back at a later time [157]. During the storage operation the optical pulse or the data signal interacts with a write pulse, which is down shifted by the Brillouin frequency of the fiber and counter propagates to the signal. Through the anti-Stokes absorption of the SBS, the data signal is depleted and the energy is transferred into a contiguous acoustic excitation. This excitation includes now the complete Information of the data signal. In this process, only a small fraction of the energy of the data signal is converted into the acoustic excitation. Most of it is transmitted to the write pulse. At a later time the data signal may be retrieved by a read pulse, which propagates in the same direction as the write pulse through the fiber. Thus, the acoustic excitation is depleted and therefore the data signal in the fiber is released again. During the process, the energy of the read pulse is transferred to the released date signal and propagates in the same direction as the original signal. Within initial experiments individual pulses and data packets with 3 bits length were stored by this method. The pulse width was 2 ns and the maximum achieved storage time 12 ns. This corresponds to a storage of 6 Bit. However, very high optical power of 100 W is required for the write and read pulses. Additionally, the respective pulse widths must be below the minimum pulse width of the data signal and the used SBS medium must be a highly nonlinear fiber. In addition, the maximum storage time is limited to the life time of the excited phonons. The exponential decay of the acoustic wave reduces the efficiency of reading the data signal until it is finally lost. The reported results show a residual amplitude of the read out data packet of only 1.8 % of the original signal with a maximum storage time of 6 Bit. On the basis of the discussed limitations this method is limited for use in the optical communication systems.

The last remaining point on the tree diagram in Fig. 5.1 is a simple method for optical data storage in optical fibers and is called **Quasi-Light Storage** [158]. It is based on the filtering of the spectrum of the data signal and can be used for nearly distortion-free [159] and variable storage of data packets in the optical fibers up to 160 ns [101].

5.1. Quasi-Light-Storage

The Quasi-Light-Storage (QLS) is a very convenient approach for the storage of data in optical communication systems. It can be easily implemented in existing fiber links, but can also be build up separately in network nodes. The main advantages include the operation over the whole transparency range of the optical fiber, especially in the C-band of optical communication, and the simple implementation with standard components from optical communications. Additionally, the QLS requires only low optical power and is independent from the modulation format or the data rate of the system. In the following sections the operation principle will be shown in detail. The experimental verification will be shown for single pulses as well as 8 Bit data patterns. Afterwards the possibilities and limitations for optical communication systems will be discussed. Finally, several methods for the storage time enhancement will be introduced and discussed.

The principle idea of QLS is based on the relationship between time and frequency domain representation of a signal [160]. A single pulse $f_{pulse}(t)$ in the time domain with the temporal width Δt, illustrated in Fig. 5.5(a), forms in the frequency range a continuous spectrum $F_{Pulse}(\nu)$ with the spectral width Δf, as shown in Fig. 5.5(b). The multiplication of the spectral representation of the pulse $F_{Pulse}(\nu)$ with an arbitrary spectral distribution $F(\nu)$ in the frequency domain leads to a convolution in the time domain and the according signals $f_{Pulse}(t)$ and $f(t)$:

$$F_{Pulse}(\nu) \times F(\nu) \;\rightarrow\; f_{Pulse}(t) * f(t) \tag{5.5}$$

The arrow denotes the inverse Fourier transformation from the frequency to the time domain and the star represents the convolution $\int_{-\infty}^{\infty} f_{Pulse}(\tau)f(t-\tau)d\tau$. A detailed description of the convolution theorem can be found in appendix A. If the pulse spectrum $F_{pulse}(\nu)$ is multiplied in the frequency domain by n equally spaced frequency components $\delta(\nu - n\Delta\nu)$ with the spacing $\Delta\nu$, called Dirac comb and symbolized by III, the result will be a sampled version of the spectral representation:

$$F_S(\nu) = \sum_{n=-\infty}^{\infty} F_{Pulse}(n\Delta\nu)\,\delta(\nu - n\Delta\nu) = F_{Pulse}(\nu)\,\frac{1}{\Delta\nu}\text{III}\left(\frac{\nu}{\Delta\nu}\right) \tag{5.6}$$

This sampling is illustrated in Fig. 5.5(d). Thereby, every frequency component of the Dirac comb is weighted with the corresponding amplitude of the pulse spectrum (see Appendix A). A multiplication in the frequency domain is equivalent to a convolution in the time domain. Accordingly, the sampled signal can be described in the time domain as:

$$f_s(t) = f_{Pulse} * \text{III}(\Delta\nu t) = \frac{1}{\Delta\nu}\sum_{n=-\infty}^{\infty} f_{Pulse}\left(t - \frac{n}{\Delta\nu}\right). \tag{5.7}$$

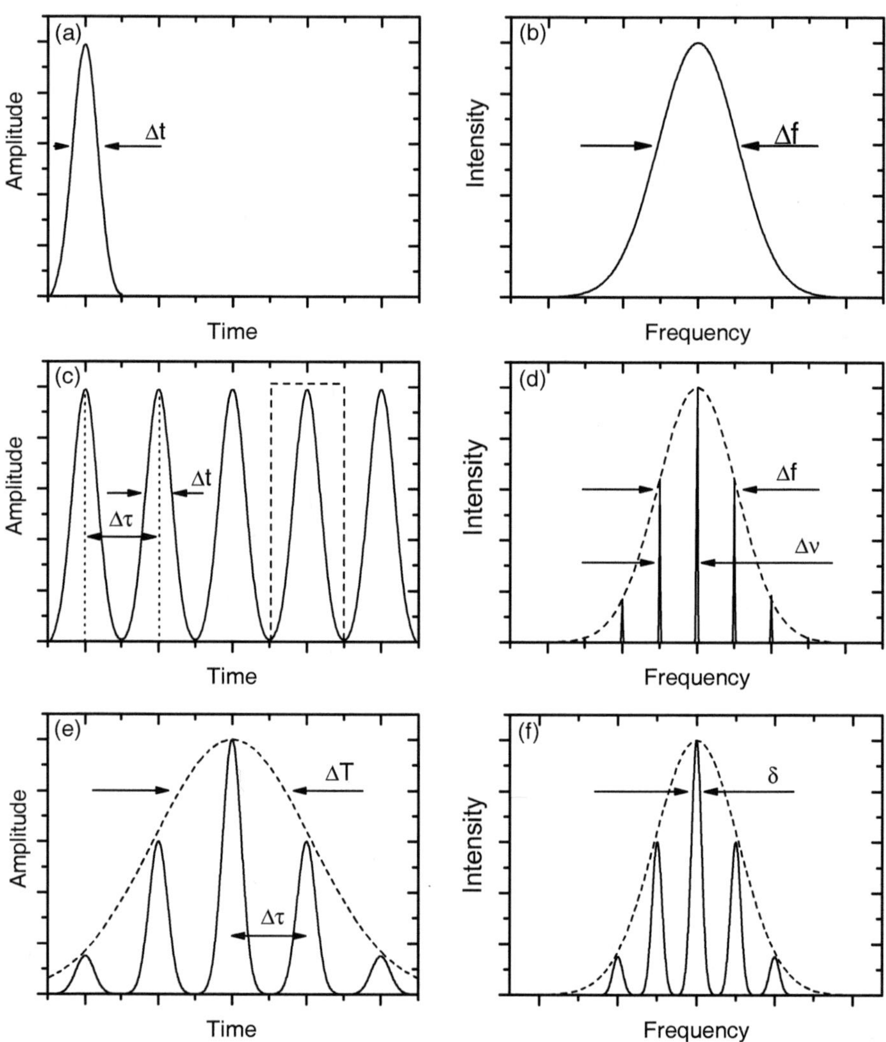

Figure 5.5.: Representation of QLS in time and frequency domain. (a) original pulse in time domain, (b) spectral representation of the pulse, (c) pulse convoluted with a frequency comb, (d) sampled spectrum by multiplication with a frequency comb. Through the utilization of SBS the sampled frequency components are broadened (f) and the time domain signal is weighted with an envelope (e).

A physical realizable system has to be causal, which means that no signal can be seen at the output before the pulse enters the input, hence $0 \leq n < \infty$. If $\Delta\tau = 1/\Delta\nu$ is higher than the temporal width of the pulse, $f_S(t)$ describes an infinite number of copies of the original pulse with the temporal distance $\Delta\tau$, as illustrated in Fig. 5.5(c)). This copying of the original pulse is the main principle of the QLS to delay or store the desired optical data. One of these time-delayed copies can be selected by an optical switch in the time domain and all other copies are discarded, which results in one delayed pulse. The switch can be easily realized by multiplying $f_S(t)$ with an appropriate rectangular windowing of the signal in the time-domain using a rectangular pulse, which is shown in Fig. 5.5(c). The delay or storage time can be adapted effortlessly fine simply by changing the frequency spacing of the individual components in the frequency comb. A larger spacing of the frequencies in the comb results in a decreased distance of the copies and vice versa. For the coarse tuning of the delay or storage time simply another copy can be selected by the time domain switch.

The theoretic model described above holds just for idealistic signals and systems. Under real circumstances, there are some facts which should be considered to explore the limitations of this method. On the one hand, the frequency comb, especially the quality of the comb, might be limited. This manifests through inconsistent amplitudes and frequency spacing, as well as, a limited number of lines and therefore limited bandwidth of the comb itself. In the latter case, the comb would not cover the whole spectral representation of the original pulse. This limitation would lead to a broadening of the pulse width after delaying and can be calculated by convoluting each pulse with the inverse Fourier transform of the envelope of the frequency comb. The next ideal assumption is that a infinitely narrow bandwidth Dirac delta function is used for sampling, which cannot be realized in a real system. Therefore, a non-zero bandwidth is assumed for each branch of the comb. Further analysis can be found in appendix A. This non-zero bandwidth leads to the creation of an envelope for the delayed pulses in time domain, ΔT, which results in finite delay times for the original pulse. As can be seen in Fig. 5.5(e), the duration of the envelope is responsible for the total duration of the input signal copies. This duration is proportional to the inverse of the bandwidth of each branch of the comb. In Fig. 5.5(f) the frequency representation of a signal sampled with a finite bandwidth is shown accordingly.

The principle setup for the realization of QLS is shown in Fig. 5.6. A Gaussian input pulse is multiplied with a frequency comb in the frequency-domain. The result in the time domain is a pulse train with equidistant copies of the input pulse. One of the pulses is extracted by a time-domain multiplication of the pulse train with a rectangular function in a modulator (Switch). For a logical '1' the signal can pass and for a logical '0' the signal is blocked. The result is a delayed, distortion-free copy of the input pulse. If the system is viewed from the outside, it forms a real memory. The input signal passes into the memory and remains there until it is released by a read signal.

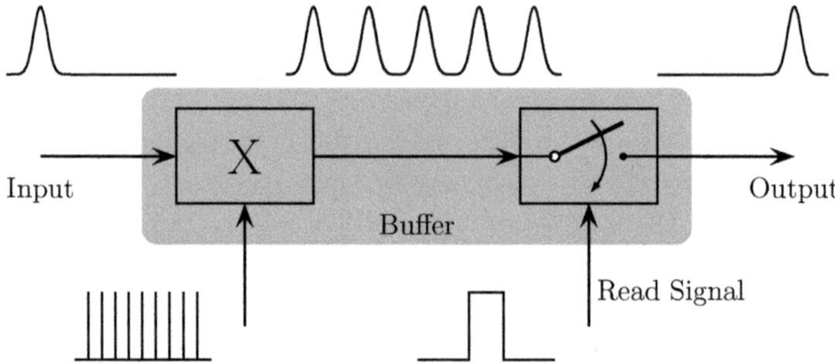

Figure 5.6.: Basic operation principle of the Quasi-Light-Storage. The X denotes the multiplication in the frequency domain.

One of the main components for the QLS is the frequency comb. During the initial experiments a simple and small frequency comb generated by a modulation of an optical carrier with a sinusoidal signal [161] was used. Later different approaches including electrical frequency mixing and arbitrary waveform generation have been investigated. In principle every other method which produces a frequency comb, four wave mixing for example, can be used as well [162]. A detailed overview about the different investigated methods can be found in appendix B. Finally, most of the experiments were carried out with a comb generated by modulation of an optical carrier with a multi tone frequency signal, which forms a Nyquist pulse train in time domain, generated by an arbitrary waveform generator (AWG). The temporal distance between the different copies $\Delta\tau = 1/\Delta\nu$ is controlled by the repetition rate of the AWG. The quality of the data storage, i.e. the distortions of the stored packet, additional noise, the output power and the maximum storage time depend directly on the quality of the frequency comb.

The next important component or process is the multiplication in frequency domain. Therefore, the nonlinear effect of stimulated Brillouin scattering in a standard single mode fiber is used (see Chap. 2). Thereby, each comb line acts as a pump wave in the sense of SBS. Thus each line generates a frequency down shifted narrowband gain, which amplify a certain part of the spectrum, as illustrated in Fig. 5.5(f). Therefore, the SBS works as a frequency-selective amplifier. This has the advantage that the delayed pulses will be amplified additionally. In principle every other periodic and narrow band optical filter can be used, e.g. Fabry-Perot filter or multiple fiber Bragg gratings. But, as discussed previously, these filter bandwidths are too broad for a useful application in QLS. Additionally, the tuning range of these filters is rather limited.

5.2. Possibilities and Limits

In optical communication systems single pulses are very rarely used. Therefore, the previous description will be extended to data packets including limitations and possible maximum storage times [163]. Analogous to Fig. 5.5 and the single pulse case, the time and frequency domain representation of a packet consisting of three bits can be seen in Fig. 5.7. For the case of simplicity, the bits and their envelope have a Gaussian shape. This has the advantage, that the spectrum shows a Gaussian shape as well. But, the QLS works for every signal function and for every signal shape. Each bit has the FWHM duration of T_{Bit} and the packet has the temporal length T_{Packet}. In the frequency domain this packet has an overall FWHM bandwidth of C/T_{Bit}. Within this envelope only separated frequency bandwidths with C/T_{Packet} are present. The constant C depends on the pulse shape. For a transform-limited Gaussian pulse $C = 0.441$ and for a rectangular function it will be $C = 0.892$ for instance [164].

If QLS is applied, the packet spectrum will be sampled in the frequency domain, as shown in Fig. 5.7(d). As a result several delayed copies of the input packet (Fig. 5.7(c)) are generated, as it is the case for the single pulse theory. A physically realizable multiplication results in broader samples in the frequency domain, defined by the Brillouin gain bandwidth $\Delta\nu_B$. The time function of the sampled spectrum is shown in Fig. 5.7(c). As can be seen, the sampling generates different undistorted and time delayed copies of the input packet. Due to the bandwidth of the sampling, the different copies are covered with an envelope. The FWHM of this envelope depends on the reciprocal of the sampling bandwidth $C/\Delta\nu_B$.

In SSMF at a wavelength of 1550nm the Brillouin bandwidth $\Delta\nu_B$ is around 30 MHz. But, for higher pump powers this can be reduced to $\Delta\nu_B \approx 10\,\text{MHz}$, as shown in Chap. 3. Therefore, the maximum storage time that can be achieved with SBS based QLS is independent of the data rate and the data format. It is just defined by the minimum SBS bandwidth and corresponds to $\Delta T_{\max} = C/\Delta\nu_B$, i.e. in SSMF it will be $\Delta T_{max} \approx 100\,\text{ns}$. In the center of the gain distribution the SBS amplification doesn't change the phase of the amplified signal [165]. Hence, the SBS based QLS is transparent for the signal modulation and can be applied to higher order modulation formats as well. The storage of binary phase shift keying (BPSK) data signal will be shown in detail later.

For a carrier wavelength of 1550 nm the Brillouin shift f_B is around 11 GHz. This means, gain and loss area will overlap and compensate for each other if the whole frequency comb

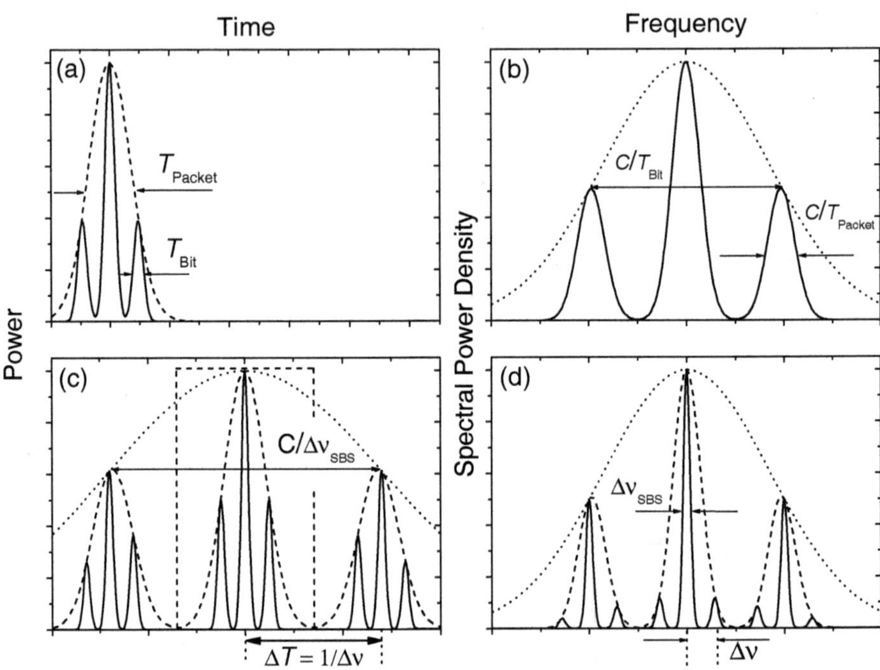

Figure 5.7.: Time and frequency domain representation of the original and sampled data packet.

extends over f_B. Hence, the maximum bandwidth of the frequency comb and therefore the data rate of the delayable data sequence is restricted by ν_B. But, this only holds for just one single pump source. If additional pump sources are included and each of which produces a frequency comb, the data rate can be simply enhanced [166]. Since every pump wave in an optical waveguide can produce its own Brillouin interaction, the SBS based QLS can be used for the independent storage of packets from different wavelength division multiplexing (WDM) channels in just one waveguide.

An optical packet of the length T_{Packet} consists of k Bit, each with the Bit length T_{Bit}. If the duty cycle is 50%, it can be written that:

$$T_{Packet} = 2k \times T_{Bit} = k/R. \tag{5.8}$$

With R as the data rate of the bits in the packet. For QLS the minimum delay between two copies has to be at least the length of the packet $\Delta T_{\min} = T_{Packet} = k/R$, otherwise an intersymbol interference will occur between different copies and the signal will be distorted. Hence, the maximum frequency difference between the frequencies in the comb is $\Delta\nu_{\max} = R/k$. Therefore, the frequency difference between the comb lines can be adjusted in the range:

$$\Delta\nu_B < \Delta\nu \leq R/k \qquad (5.9)$$

Accordingly, the storage time can be freely tuned in the range between:

$$T_{Packet} \leq \Delta T < 100\text{ns}. \qquad (5.10)$$

If smaller storage times than T_{Packet} are required, $\Delta\nu$ can be made smaller than the natural Brillouin gain bandwidth $\Delta\nu_B$. In this case the frequency comb works as a single broadened Brillouin gain and delays the pulse sequence by the group index change initiated by SBS [167], in principle a classic Slow-Light system. With this method the maximum achievable delay is around 1 Bit which can be further enhanced by several methods up to around 3-4 Bit [107], [144]. Hence, with the SBS based QLS every time delay between Zero and 100 ns with a gap between 4 Bit and T_{Packet} can be realized by a simple change of the frequency difference between the branches in the comb and by a time shift of the rectangular window function.

For a distortion free storage of the bit sequence its whole bandwidth has to be sampled and each sampled frequency must be treated in the same way. Hence, the minimum bandwidth of the frequency comb B_{CMin} is defined by the packet bandwidth. The amplitude of all comb frequencies in this bandwidth has to be the same. The bandwidth of the packet depends on the temporal width and the shape of its bits. If a duty cycle of 50% is assumed and that the bits in the packet have no sharp edges, the bandwidth of the sequence comes in the range of its data rate, thus it can be estimated that $B_{CMin} \approx B_{Packet} \approx R$. According to Eqn. 5.8, the minimum time difference between the copies was k/R. Hence, the minimum number of frequencies in the comb has to be $R/(R/k) = k$. Therefore, for QLS a flat frequency comb with a minimum bandwidth which corresponds to the data rate and a minimum number of branches which corresponds to the number of bits in the packet is required.

According to the discussion above, the QLS can at least store one packet. Therefore, the maximum number of bits which can be stored by QLS is defined by the maximum packet length. This maximum packet length is:

$$k_{\max} = R/\Delta\nu_B \qquad (5.11)$$

Thus, the capacity of the buffer (i.e. the number of bits which can be stored by QLS) increases with the data rate. For a packet with 1 Gbit/s and $\Delta\nu_B = 10\mathrm{MHz}$ this is 100 Bit and for a data rate of 100 Gbit/s it corresponds to $10,000$ Bit. Equation 5.11 can as well be seen as the delay-bandwidth product [168]. Hence, a 10 ns pulse experience a fractional storage time of 10 Bit and a 1 ps pulse of $100,000$ Bit. On the other hand, the hold-off time, i.e. the "dead time" during which the buffer cannot accept any additional data [132], is for QLS exactly the maximum storage time. So for $\Delta\nu_B = 10\,\mathrm{MHz}$, the hold-off time is 100 ns as well. The bandwidth of the SBS is defined by the convolution between the bandwidth of the natural Brillouin gain in the fiber and the bandwidth of the pump wave which produces the gain. Hence, $\Delta\nu_B$ and therefore the hold-off and the maximum storage time can be adapted to the given application by a modulation of the pump [169].

5.3. Experimental Verification

The experimental setup for the verification of the theoretical predictions can be seen in Fig. 5.8. The QLS was measured with different frequency combs shown in appendix B. During the first experiments the comb generated by electrical frequency mixing was used. During further investigation the frequency comb generated by the AWG shows the best performance for QLS since it is relatively flat and has sharp edges. Therefore, this section will just focus on the setups and results generated with this comb method. In principle, any other method which produces a flat frequency comb with sharp edges can be used.

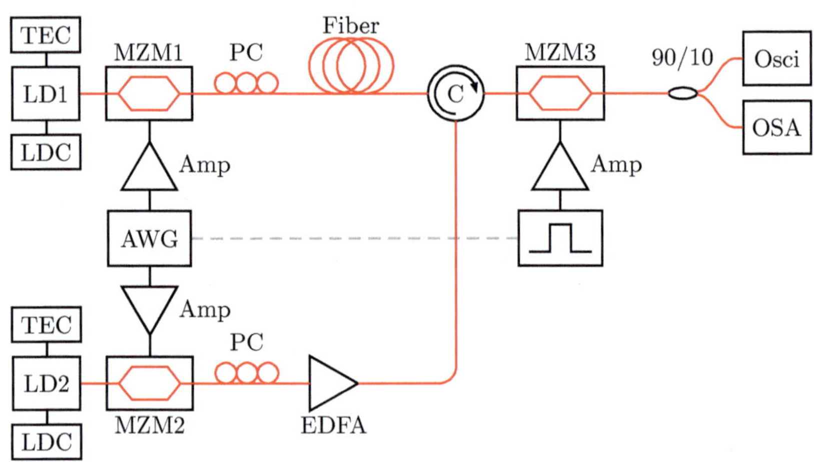

Figure 5.8.: Experimental setup for the Quasi-Light Storage.

The central element for generating all signals is the AWG. Channel one provides the data signal and the second channel the multi tone signal for the frequency comb generation. The data signal is electrically amplified (Amp) and transferred to the optical domain with the help of MZM1. The electric power of the data signal is 20 dBm and the bias of MZM1 is 4.55 V. Depending on the type of modulator used, these values vary and need to be corrected. They are adjusted in a way that the amplitude of the signals becomes maximum and the noise is minimal. In general, all modulators are equipped with polarization controllers to adjust the input polarization of the modulator and therefore the modulation efficiency, but this is not shown here. The optical carrier wave with a wavelength of $\approx$1550 nm and an output power of 10 dBm is provided by a distributed feedback laser diode (LD1). For the control and stabilization of the wavelength and the optical power of the LD, an external temperature controller (TEC) and a laser diode current controller (LDC) is used. In addition, the laser diode is protected with an isolator. The polarization for the QLS is adjusted with a PC and the data signal is coupled into the fiber. During the experiments, mainly AllWave fibers were used with a length of 20km. This leads to relatively low pump powers that are required to generate SBS. As already mentioned in Chapter 3, these fibers achieve a low Brillouin bandwidth of 10 MHz high pump powers. Thus the maximum storage time of the data packets can be 100 ns.

The multiplication is carried out in the fiber via SBS. The generation of the pump wave can be seen in the lower part of Fig. 5.8. The sinc function generated by the AWG is electrically amplified to 25 dBm and transferred to the optical domain with MZM2. The bias voltage of the modulator was 3.6 V. Therefore, the carrier is degraded in order to have the same level as the generated sidebands from the modulation. This leads to a flat frequency comb. The optical carrier wave is generated by LD2 and is also in the range of 1550 nm. The polarization of the pump wave for the QLS is also controlled by a PC. The required pump power for the SBS is provided by an EDFA. The comb is amplified to 26 dBm and coupled via a circulator (C) into the fiber. Inside the fiber the frequency comb generates multiple SBS gains and the counter propagating signal is partially amplified or sampled, respectively. Therefore, the center wavelengths of both lasers need to have a difference according to Brillouin shift. Additionally, the wavelength of the pump laser needs to be lower than the signal wavelength. The wavelengths can be easily tuned by the resistor value of the TEC. If both laser diodes are aligned correctly, the signal is sampled and the copies are generated.

These copies are coupled out after the fiber via the circulator. MZM3 acts as a time domain switch and the desired copy can be selected. Therefore, it is driven by a rectangular signal, which can be generated either by the AWG or a separate generator. However, the external generator needs to be synchronized with the AWG and an oscilloscope during the measurement. The electrical input power at the modulator was 16 dBm and the bias voltage of the modulator was 3.25 V. MZM3 is adjusted in a way that a copy of the data signal which is the rectangular modulation signal will pass and all other signals are neglected. For the analysis of the data, the signal is divided with a 90/10 coupler and investigated with an OSA and an oscilloscope. The correct positions of the LD can be monitored on the OSA. The desired extracted data packet is analyzed on the oscilloscope. A visualization of the whole experiment, including all details about the used equipment, can be found in form of a video in [170].

During the measurement different data patterns have been used. Due to the limited number of 9 lines in the used frequency comb and in order to have a small space between the delayed packets, the used length of the data pattern is 8 Bit. The generated copies of a '10101011' sequence with a data rate of 1 Gbps can be seen in Fig. 5.9. Therefore the pulse duration is 1 ns and the overall packet length is 8 ns. The used frequency comb during this measurement was generated by electrical frequency mixing. As can be seen, the copies have a distance of 10 ns. This corresponds to a frequency distance of the comb line of $\Delta\nu =100$ MHz. With the time domain switch or rather the MZM, one of the delayed copies is extracted and analyzed with the oscilloscope. The delay is almost distortion free. The changes in the amplitude can be addressed to instabilities of the system, i.e. the drift of the laser diodes. Additionally, the double ones at the end of the pattern are distorted a bit. This would mean that the lower frequency components are not sampled correctly. Therefore, the frequency comb needs to be

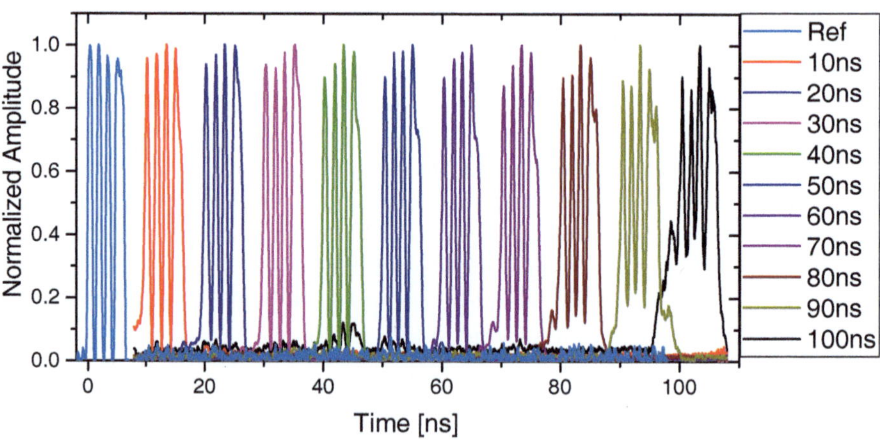

Figure 5.9.: Delayed '10101011' pattern via Quasi-Light Storage after extraction with a time domain switch.

enhanced. As can be seen, the signal to noise ratio decreases significantly for higher delays. This is a result of the lower signal power due to the envelope over the generated copies. Also a non-ideal rectangular function is used for the extraction of the different copies and might cause additional impairment to the extracted packets.

A coarse tuning is simply possible by choosing a different copy, as demonstrated in Fig. 5.9. For a fine tuning of the delay, the frequency distance between the comb lines needs to be changed. The fine tuning of the same data pattern around 30 ns can be seen in Fig 5.10. Thereby, the delay is tuned from 30 to 33 ns in 1 ns steps, which corresponds to frequency distances in the comb of 100, 96.8, 93.7 and 90.9 MHz, respectively. As can be seen, the QLS is completely tunable.

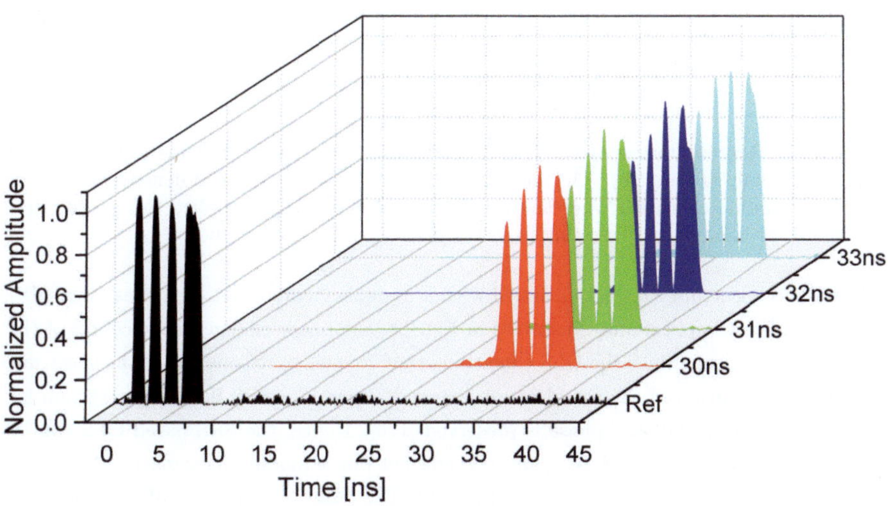

Figure 5.10.: Fine tuning of a data pattern via QLS.

5.4. Phase Modulated Signals

Advanced, spectral efficient optical modulation formats have become a key ingredient to the design of modern communication systems, because it is possible to minimize both the linear and the nonlinear impairments over the transmission fiber, e.g. chromatic dispersion and four-wave mixing. In optical communications especially phase encoded signals are becoming evermore important due to their potential for increased receiver sensitivity, tolerance to various fiber impairments and better spectral efficiency [171]. The BPSK for an '11001101' data packet can be seen in Fig. 5.11 on the left side. In the frequency domain the BPSK can be characterized by its spectral density, represented by the dashed line in Fig. 5.11 on the right side [172].

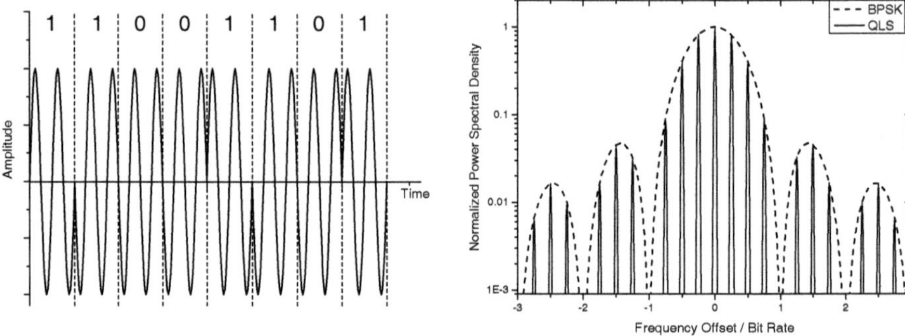

Figure 5.11.: Representation of the BPSK modulated data packet in time (left side) and frequency domain (right side). The dashed line shows the power spectral density function of the BPSK modulated packet whereas the solid line shows the QLS applied to the spectrum. The frequency components resulting from the packet length are not shown.

Again this spectrum can be multiplied with a frequency comb by the utilization of SBS. As described in Chap. 2.3 and illustrated in Fig. 2.8, the phase change in the center of the Brillouin gain is zero. Since the QLS extracts narrow frequency regions out of the spectrum, the phase information of the signal is not changed. Even in a SBS Slow-Light system for phase modulated signals the phase is preserved [173]. Hence, the SBS based QLS can be utilized for the all optical storage of spectral efficient phase modulated signals.

The experimental setup can be seen in Fig. 5.12. The basic setup for the QLS is kept. The MZM for generating the data signal is replaced by a phase modulator (PM). The PM is driven at V_π in order to achieve the best modulation results. The pump power is decreased to 20 dBm in order to reduce the distortions and especially the noise in the system. The generation of the frequency comb and the electrical data signal remains the same. The resulting copies of the QLS are combined with a LO, in form of a fiber laser, by a 3 dB coupler in order to perform coherent optical detection [110]. The polarization matching between the LO and the received signal, as well as the relative phases of the signal and the LO are ensured. The desired copy is again extracted by a time domain switch and is analyzed with an oscilloscope.

Within the storage of phase modulated signal, a 8 Bit pattern with the sequence '11001101' at a bitrate of 1 Gbps was used. The measurement results can be seen in Fig.5.13 [174]. The

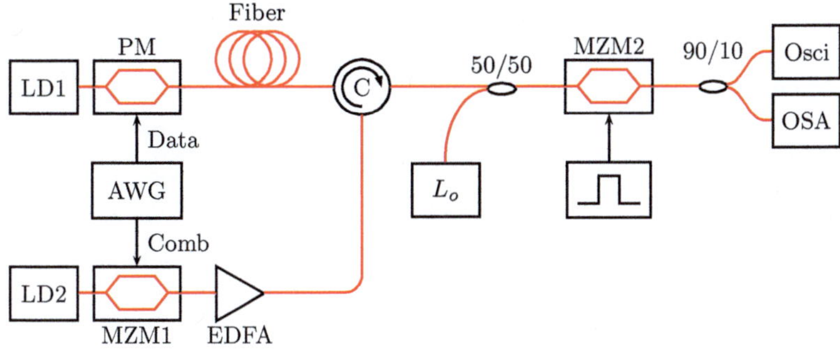

Figure 5.12.: Experimental setup for the storage of phase modulated signals.

reference pattern without QLS can be seen on the left side (black) as well as the stored packets. Due to the reduced pump power and therefore the lower Brillouin gain bandwidth, a maximum storage time of up to 60 ns was achieved. According to the theory, the delays are within an envelope. The width of the envelope is proportional to the inverse of the width of the SBS gain spectrum. The SNR, calculated as $10log(P_{signal}/P_{noise})$ with P_{signal} as the average power of the signal and P_{noise} as the average noise power, for the specific delays can be found in Fig. 5.13. Like the delays themselves, it follows the mentioned envelope.

A detailed analysis of the occurring distortions is shown in Fig. 5.14. The existing distortions can be addressed to laser drifts and changes due to the modulator over the time. Thus, the shape of the frequency comb and the power distribution of the several lines is changed. Also small variations of the polarization during the measurement can cause distortions. Another non-ideal behavior is the finite SBS bandwidth, which limits the maximum achievable storage time. During the measurement, the maximum available pump power was 20 dBm at the fiber input. Therefore the pump power was not sufficient enough to reduce the gain down to 10 MHz in an AllWave fiber. This results in a limited storage time of 60 ns. The copies after this point disappear in the noise floor. Additionally, the width of the comb was restricted to 0.9 GHz in our experiment. Therefore, not the whole bandwidth of the spectrum can be sampled correctly and some distortions occur. The frequency comb can be broadened, e.g. with an additional modulator. Since the QLS is simply a sampling in the frequency domain, with an ideal flat comb and sufficient bandwidth the QLS would be distortion free. The measurement shows that the QLS works well for BPSK modulated signals. Therefore, the results can be generalized for other spectral efficient modulation formats, like differential phase shift keying, as well. Further investigations with spectral efficient modulation formats

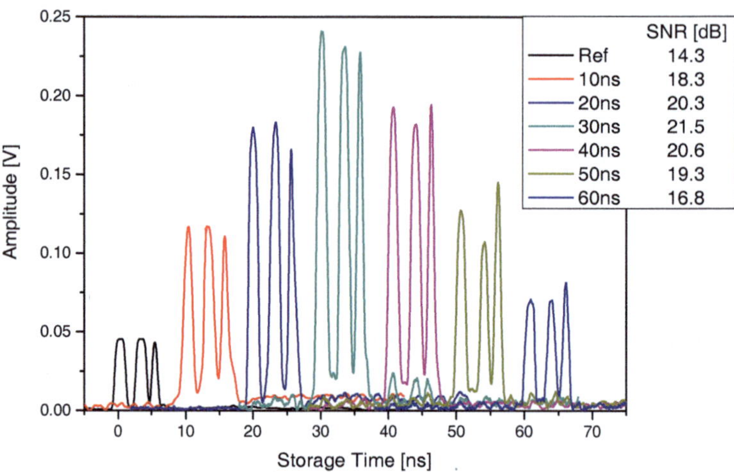

Figure 5.13.: Measurement results for a 11001101 BPSK modulated bit sequence with the reference signal on the left side (black) and the different extracted copies of the SBS based QLS.

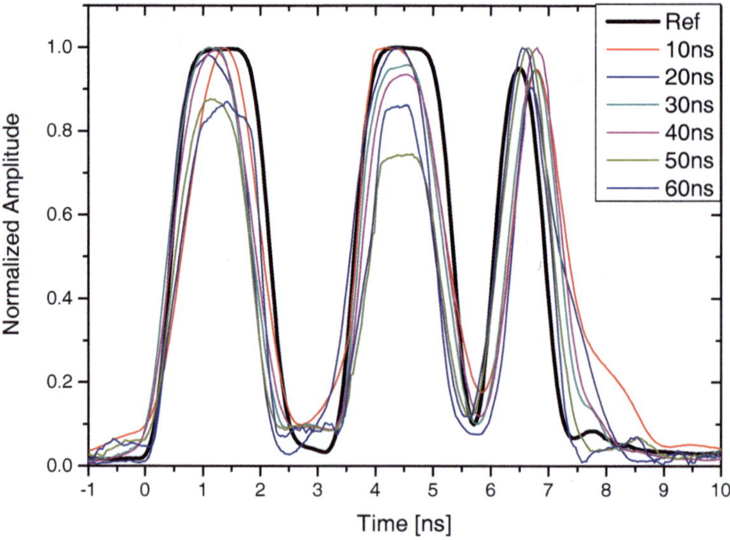

Figure 5.14.: The extracted patterns overlapped together with the reference (black line) in order to highlight the distortions.

and a mixed amplitude and phase modulation like quadrature amplitude modulation (QAM) are necessary. In conclusion the QLS show great potential for the tunable storage of optical data packets in optical communication systems, utilizing only standard telecommunication equipment.

5.5. Storage Time Enhancement

Due to the minimum SBS bandwidth of 10 MHz, which is used as filtering process within the QLS, the maximum storage time is restricted to 100 ns for a single QLS system. In order to overcome these limitations several approaches were investigated [175]. At first, the storage time of QLS can be increased by the reduction of the gain bandwidth. On the one hand, this is achieved by the superposition of the gain with two additional losses [176, 177] and on the other hand with the multi stage system [101], as discussed in Chap. 3. Secondly, the advantages of the QLS can be combined with a fiber loop to enhance the storage time [178, 179].

5.5.1. Superposition

The Brillouin gain bandwidth can be reduced significantly by the superposition of the gain with two additional losses, as described in section 3.4. This narrowing comes at the cost of a gain reduction. For the QLS this is not a disadvantage since a reduced gain means reduced noise and the signal that lies in the losses is significantly suppressed. However, the output power of the EDFA is divided by the number of lines in the frequency comb, which reduces the possible pump power for each line. Within the experiment the Brillouin gain bandwidth was reduced to 7 MHz, limited by the maximum output power of the used EDFAs. Accordingly, the maximum storage time will be $\Delta T_{max} = 1/\Delta \nu_B = 142$ ns. At this point the gain is high enough to ensure proper sampling and the storage time is enhanced by 40%. A larger enhancement of the storage time should be possible, since in theory the reduction of the gain by the losses can be compensated by higher pump powers.

The experimental setup for the QLS with reduced Brillouin gain bandwidth by the superposition of the gain with two losses can be seen in Fig. 5.15. The pump waves for the gain and the two losses are generated by LD2 on the right side. Therefore, MZM5 is driven in the suppressed carrier regime with a bias voltage of 2.73 V, driven by a sine wave at the frequency of the Brillouin shift f_B. For the used fiber this was 10.855 GHz with an electrical power of 24 dBm. The frequency comb for the QLS is again generated by a periodic sinc-function. The Fourier transform of this function is a frequency comb, as described in appendix B. The comb consists of 12 branches with a frequency spacing of 100 MHz and

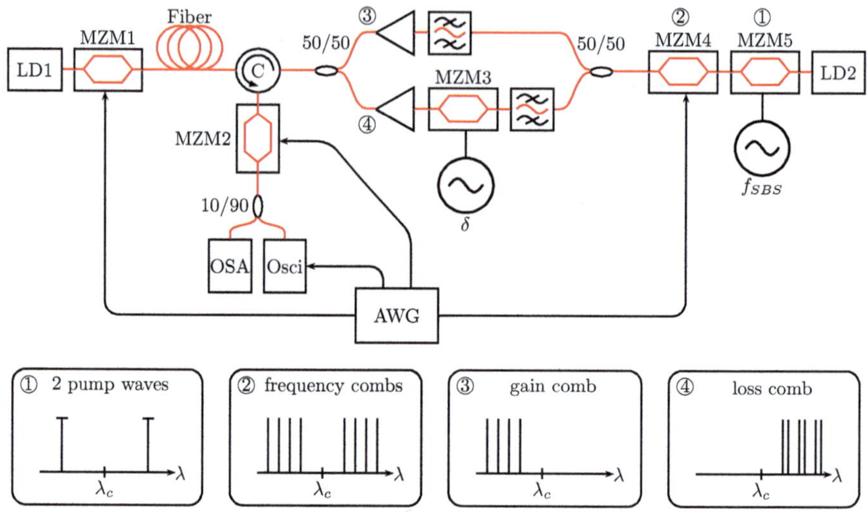

Figure 5.15.: Experimental setup for the Quasi-Light Storage with a reduced Brillouin gain through the superposition with two losses. The insets clarify the generation of the combs for the Brillouin gain and losses.

an overall bandwidth of 1.2 GHz to cover the whole spectrum of the input signal. The electrical signal with a power of 4 dBm is transferred into the optical domain by MZM4. The modulator was driven with a bias voltage of 4.84 V. Since two optical waves are coupled into the modulator, this results in two separate frequency combs. One for the Brillouin gain and one for the Brillouin loss. Afterwards the signal is split via a 3 dB coupler. In the upper path the upper sideband is filtered out with a FBG, so that the lower comb lines can be used as SBS gains. In the lower path the upper comb is used in combination with MZM3 to produce two losses for each gain. The modulator MZM3 is operated with a bias voltage of 3.38 V and is driven with a sine wave with a frequency of 15 MHz and a power of 13 dBm, which defines the distance between the losses. Afterwards, the gain and the doubled loss comb are independently amplified by an EDFA and combined via a coupler. The output power of the EDFA for the gain comb was 30 dBm and for the loss comb 27 dBm. Since the gain and the losses are produced from the same source, a fine tuning between them is very simple by the frequency of the electrical signals applied to the MZM's. The pump waves are coupled into the fiber via a circulator, where several Brillouin gains are generated by the one comb and several doubled Brillouin losses for the other comb.

Within the experiment an 8 Bit data packet with a data rate of 1.15 GB/s was used. The packet is again generated with an AWG. The optical carrier is provided by LD1 and the electrical signal is transferred into the optical domain by MZM1. The electrical signal power was 24 dBm and the bias voltage of the modulator 4 V. The data packet is coupled into a 20 km AllWave fiber. The time domain switch is realized by MZM2, where one of the copies generated by QLS is extracted. This modulator is driven by a rectangular signal with a power of 4 dBm, which is also generated by the AWG. The bias voltage was set to 3.65 V in order to ensure extraction of the desired pulse and discarding of the unwanted copies. The extracted signal is split with a 90/10 coupler and then detected with an oscilloscope which is triggered by the AWG to analyze the data. The optical spectrum is monitored with an OSA to ensure the correct alignment of the lasers.

The used reference data pattern before the QLS system can be seen in Fig. 5.16(a). The maximum achieved storage time, with the superposition of the gain with two losses, was 140 ns, as can be seen in Fig. 5.16(b). The different copies generated by the QLS are shown in Fig. 5.16(c). The stored pattern shows very low distortions. For the higher storage times the amplitude of the extracted copies is rather low. Hence, in Fig. 5.16(c) the noise seems to be increased through the normalization.

In Fig. 5.17(a) the different stored packets up to 140 ns together with the reference are shown. The delays are overlapped in order to highlight the distortions which are caused by the system. The stored versions show a slight broadening and some power fluctuations. This can be addressed to the insufficiency of the frequency comb as well as fluctuations during the measurement. For a distortion free sampling, the frequency comb has to be flat over the whole bandwidth of the signal. The used frequency comb showed some fluctuations of around 1 dB between the different frequency components. Therefore the signal is not sampled equally. Also a non ideal rectangular function is used for the extraction of the desired copy.

The relative pulse width versus the storage time can be seen in Fig. 5.17(b). Therefore, the full width at half maximum pulse width of the first '1' of each stored data packet is considered. The reference pulse had a width of 0.87 ns and the maximum broadening occurred with 0.94 ns for the 130 ns pulse. This equals a maximum broadening of just 8% of the original pulse. As can be seen in Fig. 5.17(b) the pulse width for the first three stored pattern increases and then, for the next three there is almost no pulse broadening. This can be addressed to some fluctuations of the comb during the measurement. In addition, the power fluctuations are very small and at no time larger than 20%.

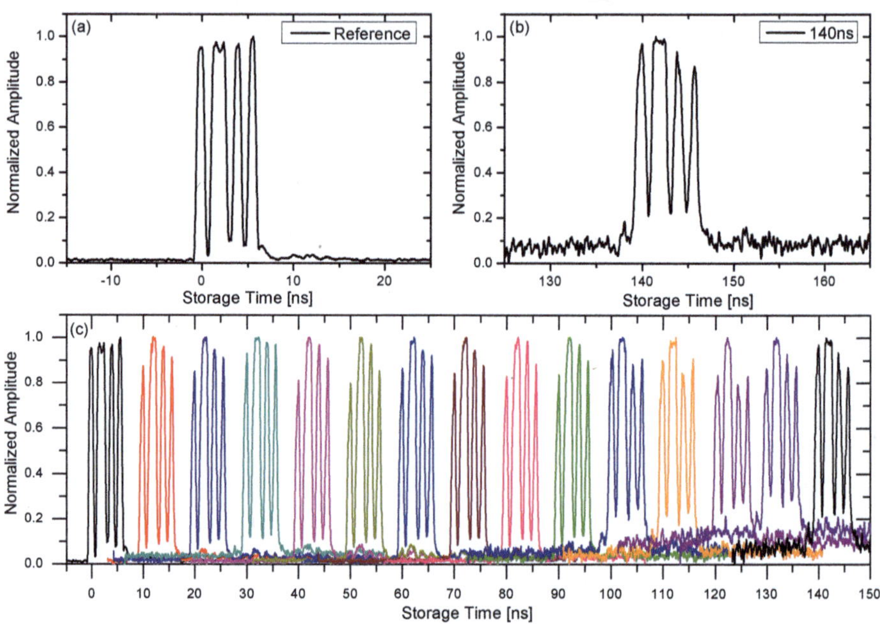

Figure 5.16.: Measurement results for the QLS with reduced Brillouin gain bandwidth by superposition with two losses: (a) reference pattern, (b) 140 ns stored data pattern and (c) train of copies.

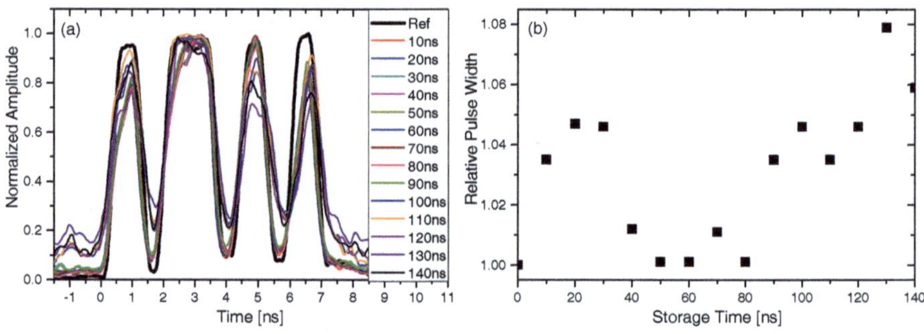

Figure 5.17.: Distortion analysis of the delayed and extracted data packets after the QLS system (a) and the measured relative pulse width versus the storage time (b).

5.5.2. Multi stage system

As mentioned previously, some methods for the Brillouin gain bandwidth reduction show disadvantages for the utilization within QLS. Within the superposition of the gain and loss the power of the EDFA is spread to all lines in the combs, and especially in the loss comb, resulting in limited pump power for the system and therefore limited bandwidth reduction. Another approach that can be utilized for the storage time enhancement of QLS is the multi stage SBS system. With this method, the storage time of QLS can be easily enhanced without any of the mentioned disadvantages. As discussed in Chap. 3.2 the Brillouin gain bandwidth can be reduced down to 5.8 MHz in a three stage system. Accordingly, the maximum storage time will be $\Delta T_{max} = 1/\Delta\nu_B = 172$ ns. Basically, the experimental setup is a combination of the previous ones and can be seen in Fig. 5.18. A detailed description of the multi stage block can be found in section 3.2. The data signal, as well as, the comb is generated electrically with the AWG. Two independent laser sources provide the optical carriers and the signals are transformed into the optical domain with the help of MZM's. The according power and bias values can be found in the previous sections. The signal as well as the comb is coupled into the multi stage system, represented by the three framed SBS blocks. The detailed setup for the multi stage system can be seen in Fig. 3.5. The pump power for every stage was 26 dBm and the input signal power to the multi stage system was 0 dBm. The generated copies are coupled out of the system and one of them is extracted through MZM3 driven by a rectangular signal. The extracted signal is analyzed with an oscilloscope and the alignment of the lasers is monitored with the OSA.

In order to show the delay of different data rates and bit pattern lengths, a 4 bit sequence with a data rate of 0.5 Gbps was used. The generated copies by the QLS can be seen in

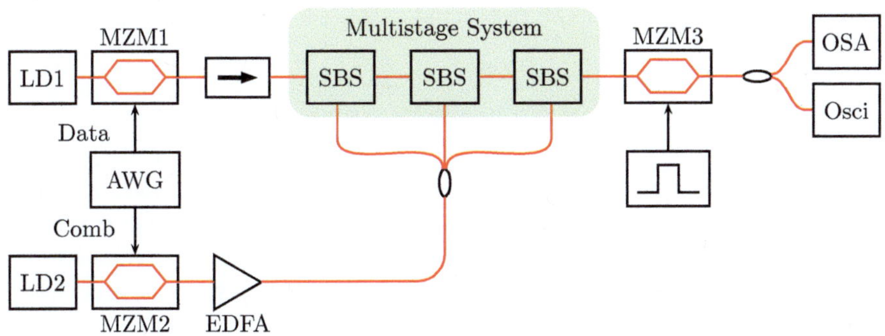

Figure 5.18.: Experimental setup for the Quasi-Light Storage in a multi stage system.

Fig. 5.19. Every delay was extracted separately from the generated copies. The maximum storage time of the QLS could be enhanced to 160 ns. Theoretically, a maximum storage time of 170 ns should be possible at a Brillouin bandwidth of 5.8 MHz. However, due to the limited output power of the EDFAs the measurement is restricted to 160 ns. This still equals an enhancement by 60% compared to the original QLS. The results show rather low distortions, which can be addressed to the envelope of the QLS itself as well as the insufficiency of the frequency comb. But, for higher delays the signal to noise ratio significantly increases.

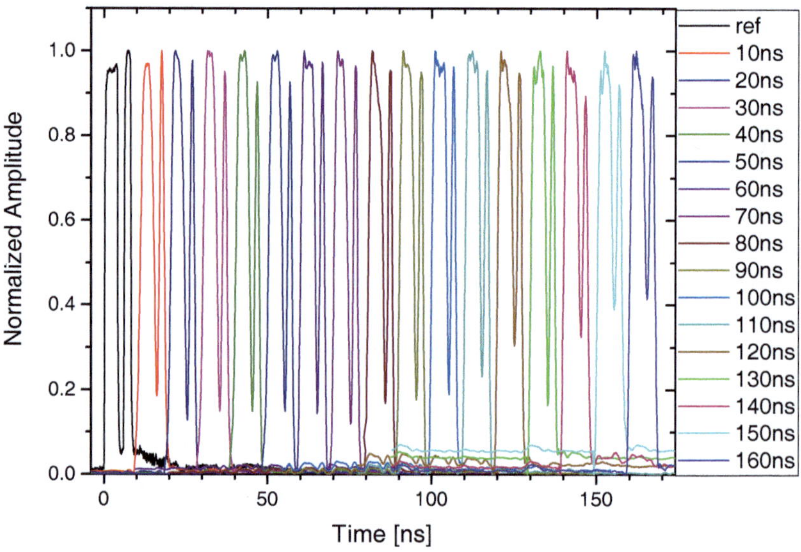

Figure 5.19.: Measurement results for the delay of a '1101' bit sequence in a multi stage QLS system.

5.5.3. Feedback

Another possibility for the enhancement of the storage time is the combination of the QLS with a fiber loop. The fiber loop itself can lead to very high storage times, but this storage time is not tunable and is predefined by its length. This issue can be overcome by the utilization of QLS. The basic operation principle can be seen in Fig. 5.20. The data signal is coupled into the system and several copies of the signal are generated due to the convolution with the frequency comb, as described in the previous sections. One of these copies is extracted and fed back into the system. In order to avoid saturation of the SBS gain in the fiber, the signal is attenuated before every new round trip. Since QLS is carried out again, for the next round again a maximum tunable storage time of 100 ns is possible. This can be carried on until the SNR becomes too low. Within a first proof of concept experiment 10 round trips with a maximum tunable storage time of 500 ns were achieved. The number of round trips during the experiment was not restricted by the SNR, but rather by trigger issues of the used configuration of devices. Hence, much higher storage times should be possible. It has been shown that the maximum number of round trip times in a fiber loop can be as high as 100 times [155]. With an average storage time of 50-100 ns per round this would lead to a maximum storage time of 5-10 μs. However, in [155] the number of rounds was restricted by the noise induced by the amplification of the signal at each round.

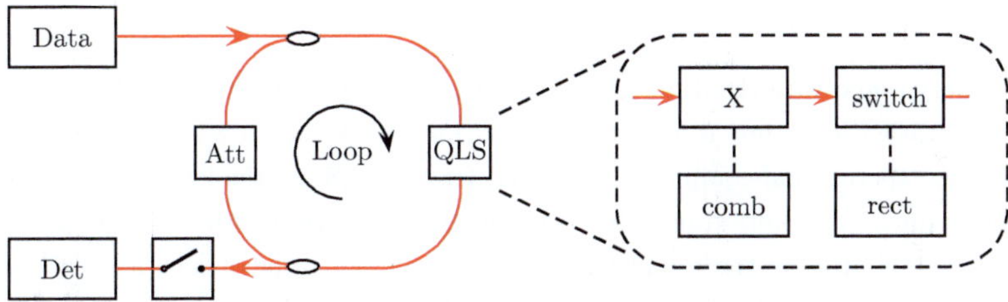

Figure 5.20.: Principle setup of the QLS within a loop. The X denotes the multiplication of the signal with the frequency comb.

The experimental setup can be seen in Fig. 5.21. The data signal with a length of 8 bit and a bit rate of 1 Gbps was generated by channel 1 of the AWG with an output power of 4 dBm and transferred into the optical domain by LD1 and MZM1. The optical output power of the LD was 10 dBm and the bias voltage of the MZM was 2.39 V. The data signal is repeated every 2 ms by the AWG. This signal is coupled into the 25 km SSMF, where the multiplication between data and comb takes place. To avoid damage of the equipment an isolator is used in front of the fiber. On the right side of Fig. 5.21, the frequency comb is generated by channel 2 of the AWG as described in appendix B. The comb signal is transferred into the optical domain by LD2 and MZM2. The output power of the LD was 10 dBm and the bias voltage of the MZM was 3.75 V to ensure a flat comb. Afterwards the comb is amplified by an EDFA to an output power of 24 dBm to provide sufficient pump power for the system and coupled into the fiber via a circulator. With help of MZM3, one of the copies, generated by the QLS, is extracted. This modulator is driven by a rectangular signal which is also generated by the AWG and a bias voltage of 3.25 V. This signal defines the delay for every round. The width of the rectangular window is 10 ns. The overall extracting signal consists of a rectangular

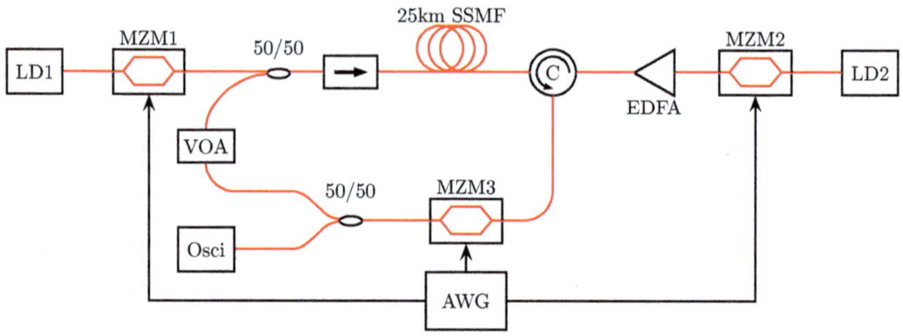

Figure 5.21.: Experimental setup for the QLS in a loop configuration.

pulse for every round with the distance of the propagation time through the system, which is $120\mu s$. This rather long time is a result of the 25 km SSMF, which was used to reduce the efforts to the experimental equipment. However, for the QLS not the length of the fiber but the transfer function is important. Recently, the SBS on a chip with a length of 7 cm has been shown [180]. Hence, the round trip time and the foot print of such a storage device can be drastically reduced. The extracted copy of the original data pattern is split via a 3 dB coupler. One part is detected with an oscilloscope, which is triggered by the AWG. The other part is fed back into the system with an additional attenuation of 20 dB in order to avoid saturation and QLS is carried out again. To investigate the different rounds, the selected copy of every different round appears at the output. For applications the system has to choose one specific round at the output. Therefore just one additional modulator will be needed.

During the first proof of concept experiments just the variable storage time of QLS is investigated and the propagation time through the fiber neglected. The variable storage time is defined by the QLS. A storage time of 0 ns means that the original pattern passes through all the rounds without the QLS. If QLS is turned on, it is possible to set up variable storage times from 0 ns up to 100 ns for every round. The overall storage time is the summation of the storage times of all round trips. Within the experiment an 8 Bit data packet with the sequence 10010011 and a data rate of 1 Gbps was used. From the generated train of copies by QLS, the copy with the highest amplitude at 50 ns is extracted in every round and fed back into the system. The reference pattern can be seen in Fig. 5.22(a) and the stored versions of the signal in the different rounds are shown in Fig. 5.22(c). A detailed view of the extracted pattern after 450 ns can be found in Fig. 5.22(b). Within the first experiments a maximum storage time of 500 ns with modest distortions could be achieved, which equals 10

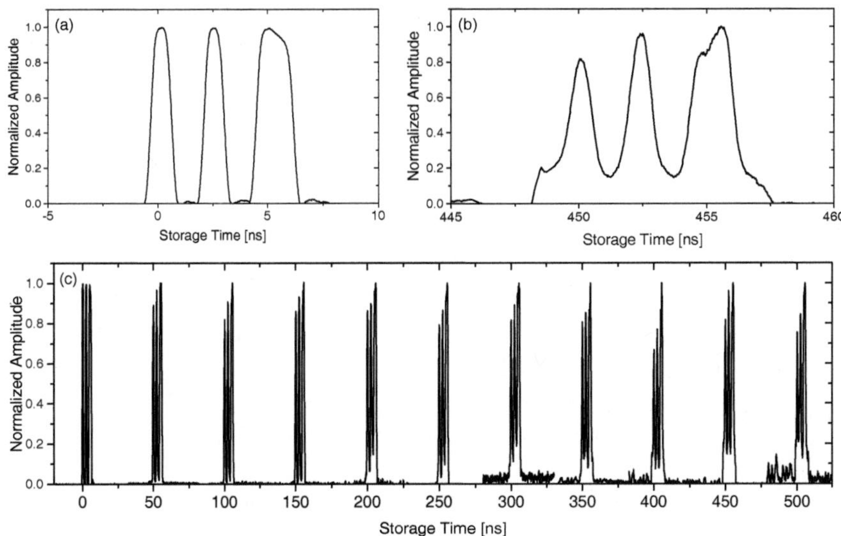

Figure 5.22.: The reference pattern on the left side (a), the extracted pattern after 450ns on the right side (b) and the stored pattern of each loop at the bottom (c).

round trip times in the system. The stored versions show a slight broadening and some power fluctuations. This can be addressed to the insufficiency of the frequency comb. A comb with a higher bandwidth would reduce this broadening because in theory a distortion free storage is possible with frequency sampling. Also, for a correct sampling, the frequency comb has to be flat over the whole bandwidth of the signal. The frequency comb we used had some fluctuations of around 2dB between the different frequency components and varies over the measurement time. Therefore the signal is not sampled equally. Also a non ideal rectangular function is used to extract the desired copies.

For the first round the additional attenuation in the loop has to be very high in order to avoid saturation. But the amplitude of the signal in the following rounds decreases. Hence, the additional attenuation is too high and the amplification due to the SBS can not compensate this. Due to the decreasing amplitude of the signal the experiment was limited to 10 rounds. To overcome this, the pump power or the additional attenuation in the loop should be adapted separately for every round by applying a time dependent function to the VOA and/or EDFA. The decreasing amplitude and not the SBS amplification is the main reason why the noise seems to be higher for the last copies. Therefore, much higher storage times in the micro- or maybe millisecond range would be possible with this method. However, this requires further investigations.

Frequency Comb Processing

By definition, frequency combs consist of equally-spaced spectral lines over broad wavelength ranges. During the last decades the generation of frequency combs was fast paced, based on the growing importance for multiple applications [181]. The development of optical frequency combs revolutionized optical frequency metrology [182, 183], high precision spectroscopy of materials [184, 185] and the development of high accuracy optical clocks [186, 187].

The generation of frequency combs is diverse. With the parametric process of cascaded four-wave mixing [188] in combination with monolithic microring resonators [189] large octave spanning frequency combs with repetition rates of 220 GHz and 100 GHz, respectively, can be generated. Another and simple way to generate a frequency comb with variable spacing is based on electro-optical modulation of a radio-frequency signal, where multiple sidebands of a carrier are produced. Frequency combs with a bandwidth of 230 GHz and a frequency spacing of 10 GHz were presented [190]. This approach was further developed by cascading phase or intensity modulators, or a combination of both [191]. However, the overall bandwidth of these combs is rather limited. A more compact comb generator is a Fabry-Perot Electro-Optic modulator [192]. Due to the specific cavity, the phases of the different frequencies are locked. They are capable of generating combs with a span of more than 10 THz. However, the modulation frequency for these devices as well as the wavelength of the laser source needs to fit the cavity length or resonance, respectively.

A typical source, that generate a large number of equidistant and phase locked frequencies, is a mode locked laser (MLL) [164]. The time domain output of these devices is typically a train of short pulses. Especially, fiber lasers that generate femtosecond pulses can produce frequency combs with a very high quality and bandwidth. Accordingly, a pulse with the temporal width of 80 fs would equal a spectral distribution of equidistant lines over 12.5 THz. At the same time, these femtosecond fiber lasers have a small footprint, are cost effective, and have a center frequency in the range of the C-band of optical telecommunications. Typical devices have repetition rates in the range of 100 MHz and a bandwidth of 15 THz. A further spectral broadening by self-phase modulation and four wave mixing in a nonlinear element like an optical fiber can enhance the spectral width to 100 THz and more. Exemplary, the output spectrum of the used fs-laser FemtoFErb 1560 can be seen in Fig. 4.21(a) in section 4.3, including a detailed measurement of the repetition rate with a BOSALO in Fig. 4.21(b).

In principle, the frequency spacing of most of these frequency combs is far too narrow for conventional optical filters, as described in section 2.6. Therefore, the narrowband characteristics of stimulated Brillouin scattering are predestined for the filtering and processing of frequency combs [193]. Especially, the polarization pulling effect enables the extraction of single lines including the suppression of all out-of band components.

Depending on the number of filtered lines, different applications can be realized. If just one of the comb lines is extracted, as illustrated in Fig. 6.1, it can be used as a narrow linewidth, tunable laser source [194]. The gray lines depict the original frequency comb lines, the red line represent the single extracted comb component, and the small black lines represent the suppression of the out-of-band components due to polarization pulling assisted stimulated Brillouin scattering. With an appropriate modulation of the extracted line, it can be long-term stabilized in the Hz region and additionally fine tuned [194]. The coarse tuning is enabled by selecting a different line; thus, the broad bandwidth of the comb and the possibility of nonlinear broadening, in principle, mean that a tunability of more than 100 nm could be achieved.

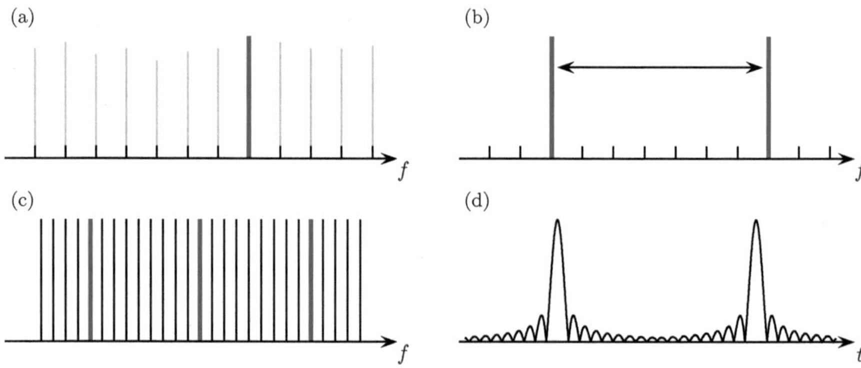

Figure 6.1.: Illustration of the extraction of single comb lines out of a frequency comb by utilizing PPA-SBS. The applications include a single line laser (a), mm-wave generation (b), the creation of a flat rectangular frequency comb (c) and consequently the generation of almost ideal sinc-shaped Nyquist pulse trains (d).

Beyond this, the selection of two lines out of the frequency comb leads to very narrow, stable and tunable millimeter or THz-waves [195, 196], as displayed in Fig. 6.1(b). The generated frequency can be easily tuned by the selection of different lines from the frequency comb. First measurements have shown linewidths in the lower Hz range and a phase noise below -104 dBc/Hz [195].

Additionally, the extraction and further processing by two coupled modulators of one or more lines can lead to perfectly rectangular shaped frequency combs with almost no power fluctuations, where the bandwidth and repetition rate can be tuned easily [197]. Linked by the time-frequency coherence this frequency combs can lead to very short, high-quality Nyquist pulses with an almost ideal sinc-shape, as long as the phases of all lines are locked to each other [198]. The extraction of three lines is schematically shown in Fig. 6.1(c) by the red lines. A subsequent external modulation of these lines leads to very flat and rectangular shaped frequency combs, as illustrated in Fig. 6.1(c). The time domain representation forms an ideal sinc-shaped Nyquist pulse train, as depicted in Fig. 6.1(d). First experiments have shown flat, rectangular frequency combs with a maximum bandwidth of 260 GHz with 27 equidistant frequency components and a power deviation of only 0.6 dB [197]. Additionally, almost ideal sinc-shaped Nyquist pulses with a pulse width of 3.5 ps and a duty cycle of 2.2% have been reported [198]. A detailed description of the mm- and THz-wave as well as the Nyquist pulse train generation will be given in the following sections.

6.1. THz- and mm-Wave Generation

During the last years, there has been a great research effort in the development of suitable technologies for the generation of millimeter (30–300 GHz) and THz-waves (0.3–30 THz) [199] due to an increasing number of possible applications like imaging for security [200], biomedicine [201], spectroscopy [202] or short-range high-data rate communications [120], to name just a few. With increasing data rates, more bit per symbol need to be encoded. Therefore, higher order modulation formats, that encode the information in phase and amplitude, are incorporated [203]. The requirements regarding linewidth and phase noise of the carrier wave increase significantly with the baud rate and the number of bits per symbol, since a certain level of phase noise is more critical for closer phase distances. All of the above applications require stable, and often tunable, sources of continuous-wave THz-frequency radiation, having a narrow linewidth and low phase noise. In principle, the generation of mm- and THz-waves can be divided in two domains. The first one is based on the electrical domain. Therefore, the limits of electronic devices like Resonant Tunneling Diodes [204] are enhanced in order to reach the THz-range. Additionally, millimeter- and THz-waves can be generated by electronic up-conversion of radio and microwave-frequency tones [205, 206]. Nevertheless, the noise associated with electronic up-conversion scales quadratically with the harmonic frequency-multiplication order, whereas the power decreases.

The second approach is based on the optical domain, where the mm- and THz-waves are mainly generated by optical heterodyning. The same process is used for optical spectrum analysis, as described in chapter 4 through Eqn. 4.3. During the heterodyne process, two waves are mixed in a nonlinear element like a photo diode, as can be seen in Fig. 6.2. The output frequency of the photo diode will be given by $\omega_2 - \omega_1 = \Delta f$ and represents the difference frequency. The two required optical waves can be provided by two different laser sources or by a single optical source. When two independent laser sources are used, continuous tunability is feasible, but due to the uncorrelated optical waves, the minimum linewidth of the generated wave can be only in the hundred kHz-range [207]. In the case of one source, a dual-mode structure with low phase noise can be used. Though the tunability is very limited and the achieved linewidths are in the MHz range [208].

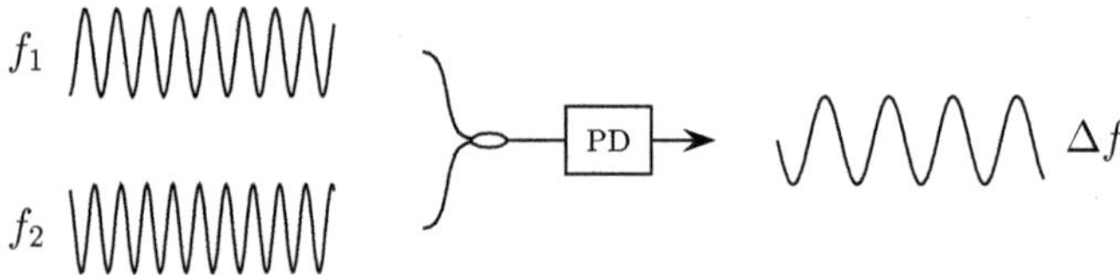

Figure 6.2.: Operation principle of heterodyne mixing for the generation of mm- and THz-waves.

For the generation of high quality mm- and THz-waves with low phase noise and ultra narrow linewidth, the two optical waves need to have a fixed phase relation to each other. This condition is fulfilled by optical frequency combs generated by mode locked fiber lasers. For the separation of the two different comb lines again the polarization pulling assisted SBS is used. This has the advantage that all remaining and unwanted components of the frequency comb can be easily suppressed.

The experimental setup can be seen in Fig. 6.3. The frequency comb is generated by a MLL with a repetition rate of 100 MHZ and an overall output power 17 dBm. A measurement of the used frequency comb can be seen in Fig. 4.21(a) in section 4.3. In order to increase the power of each line in the comb a bit, the spectrum is prefiltered by a WS and amplified with a pre-amplifier. According to the theory for the polarization pulling assisted Brillouin process, the polarization of the MLL is set to a state where it is blocked at the PBS with help of the polarization controller (PC1). Afterwards, the signal is coupled into a 20 km long

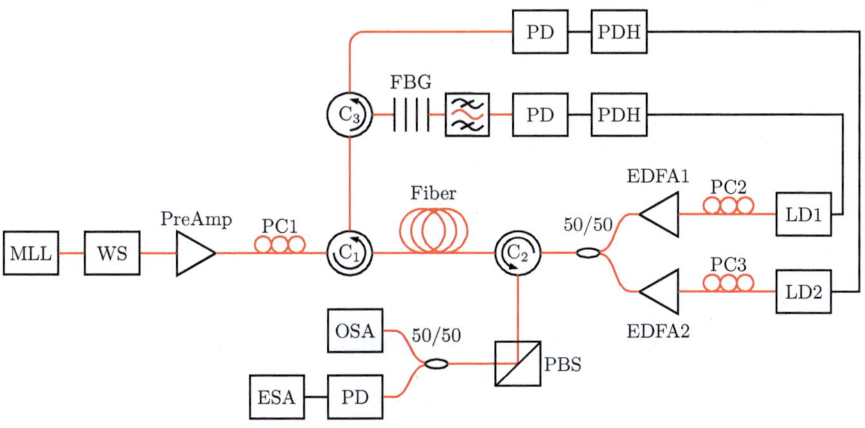

Figure 6.3.: Experimental setup for the generation of high quality mm- and THz-waves.

AllWave fiber through a circulator (C1). For the extraction of two lines out of the comb also two pump waves are required. They are generated by DFB laser diodes and amplified separately by an EDFA. The output power of both EDFAs was 15 dBm. Afterwards they are combined by a 3 dB coupler and injected into the fiber with circulator C2. The pump waves are counter propagating to the spectrum of the MLL.

Due to a temperature drift of the MLL the frequency of each of the two DFB lasers is automatically adjusted to provide continuous amplification of a particular comb line. This is realized with help of Pound-Drever-Hall (PDH) modules [209]. Therefore, the two pump laser diodes are modulated with a reference frequency. After passing the fiber, the pump waves are coupled out by the circulator C1. The pump waves are separated by a FBG that reflects one particular wave, which is coupled out by C3, and transmit the rest to a second filter that selects the other wave. During the SBS process the power from the pump waves is transferred to the amplified frequency comb lines. This pump depletion can be measured with a photo diode. The PDH-modules utilize the depletion of the pump waves for the control of the laser current of LD1 and LD2 in order to shift them and ensure continuous amplification of the desired lines. The two amplified comb lines are coupled out of the fiber by C2 and pass through a PBS. Afterwards the signal is split by a 3 dB coupler and detected by a broadband photo diode, as well as an OSA and ESA. The photo diode had a 3-dB bandwidth of 100 GHz. In order to access the frequency range at the ESA, it is used in combination with external frequency mixers. With the OSA the position of the lasers and the filters was monitored.

The SOP of the pump waves is controlled with the polarization controllers PC2 and PC3.

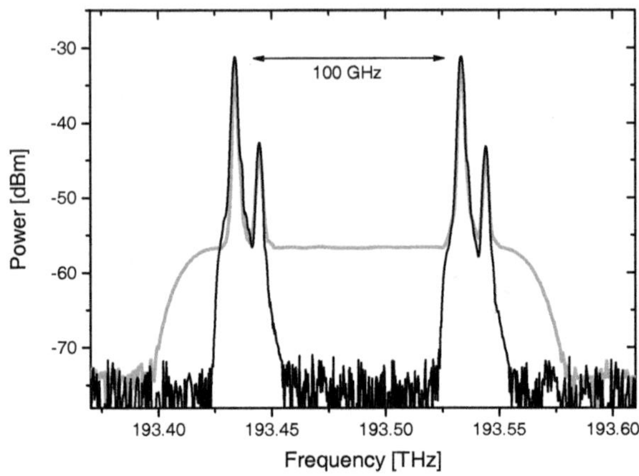

Figure 6.4.: Amplified lines of the MLL with (black) and without (gray) polarization pulling assisted SBS.

They were aligned so that the states of polarization of both pump waves are the same and partially pass the PBS. PC1 was adjusted in a way that all unamplified comb lines are blocked-off entirely by the PBS, whereas the two spectral lines of interest are partially transmitted, due to the SBS polarization pulling as discussed in section 2.4 and experimentally shown in section 4.2. The suppression of the remaining and unwanted comb lines are shown in Fig. 6.4. Exemplary two lines with a distance of 100 GHz were selected. The gray line specifies the prefiltered part of the frequency comb and the amplification by SBS but without the PBS. The black curve shows the polarization assisted SBS with the PBS in the system. As can be seen, the unwanted comb lines are additionally suppressed. The two additional peaks, around 10 dB lower than the amplified sideband and up-shifted in frequency by around 11 GHz, are the Rayleigh back scattered pump waves. If required, the power of the extracted lines can be enhanced either by higher pump powers or by an additional amplification with an EDFA.

During the measurement several mm- and THz-waves were generated. Figure 6.5(a) shows the measurement of the electrical power spectral density of the beat signal that was generated by mixing two selected frequencies that are spaced by 24 GHz. The FWHM linewidth of the generated mm-wave, directly measured with an ESA, was <1 Hz, limited by the resolution bandwidth of the ESA of 1 Hz. During the measurement the video bandwidth of the device was set also to 1 Hz with a measurement span of 100 Hz. The according phase noise measurements were carried out with the ESA as well and can be seen in Fig. 6.5(b). The phase noise at a

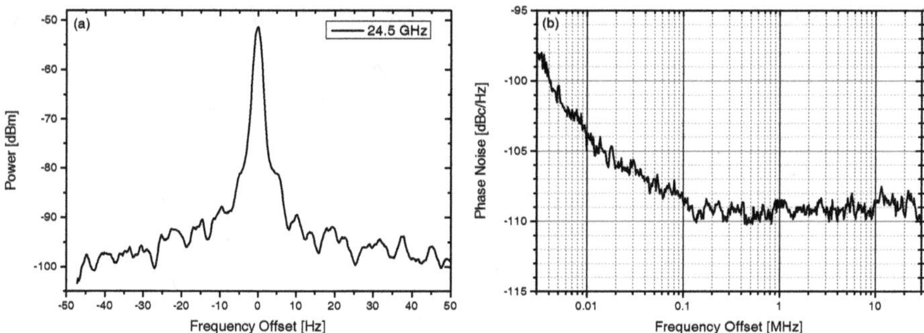

Figure 6.5.: Generated microwave signal (a) and the according phase noise measurement (b).

frequency offset of 10 kHz from the optically generated micro-wave carrier was -104 dBc/Hz. The obtained phase noise is several orders of magnitude lower than in the best reported results for setups were two external lasers were locked to two comb lines [210]. In the given setup the DFB laser diodes are locked to the frequency comb as well through the PDH-modules. For comparison the two DFB were heterodyned at the output of the first circulator. The measured linewidth was 3 MHz and the according phase noise was -67 dBc/Hz. Compared to the results achieved with the lines of the frequency comb, the linewidth is at least 6 orders of magnitude broader and the phase noise is 4 orders of magnitude larger.

The characterization of waveforms exceeding 25 GHz required additional mixers at the electrical spectrum analyzer, resulting in a reduced resolution bandwidth of 300 Hz. The

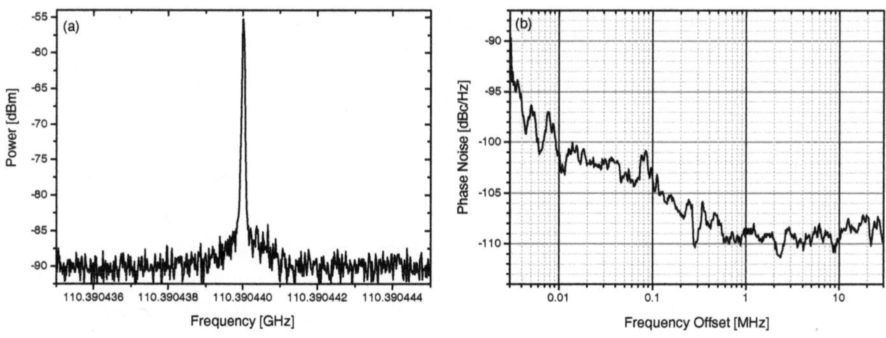

Figure 6.6.: Linewidth measurement (a) of a generated mm-wave signal with a frequency of 110 GHz and the according phase noise measurement (b).

measurement of the generated mm-wave signal with a frequency of 110 GHz can be seen in
Fig. 6.6(a). The measured linewidth is <300 Hz, limited by the resolution bandwidth of 300 Hz
of the ESA during the use of the external mixer. The according phase noise measurement is
shown in Fig. 6.6(b). The phase noise was -101 dBc/Hz at 10 kHz offset. This lower phase
noise might be related to the imperfections of the setup and the additional mixer.

The generated frequency can be tuned variable by selecting different lines out of the
frequency comb. Exemplary, the selection of lines with different distances can be seen in
Fig. 6.7. With an appropriate mixer, frequencies of 1 THz (black), 2 THz (red) and 3 THz
(blue) would be generated. The suppression of unamplified tones is somewhat degraded with
increased frequency separation, due to polarization mode dispersion in the Brillouin gain
medium.

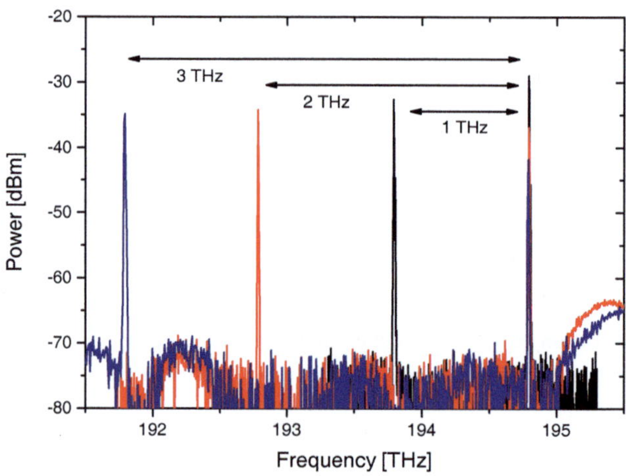

Figure 6.7.: Selective amplification of two comb lines with different spacing using polarization
pulling assisted SBS.

The generation and transmission of THz-waves with this method has been demonstrated
in [195]. For the heterodyne process and the emission of the THz-frequencies photo mixers
based on InGaAs on InP were used [211]. Therefore the setup in Fig. 6.3 needs to be adapted
as shown in Fig. 6.8. Basically, the components after the PBS are changed. The EDFAs
are used to provide sufficient input power to the photo mixers. The THz-transmitter (Tx)
and receiver (Rx) are taken from a commercial spectroscopy system from Toptica Photonics
AG. The two extracted lines out of the comb are mixed at the Tx and transmitted over a
distances 24.5 cm. Since the output power of these mixers is just in the range of some μW,

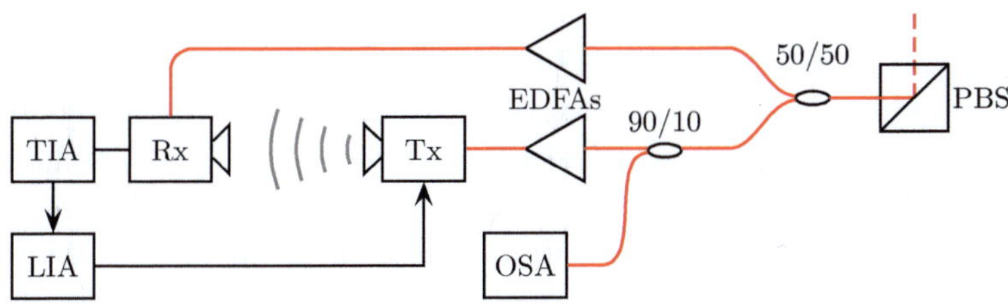

Figure 6.8.: Additional setup for the generation and transmission of mm- and THz-waves.

the maximum transmission distance is rather limited. At the Rx the signal is down converted and measured with a transimpedance amplifier and a lock-in amplifier. Therefore, the signal is modulated by chopping the Tx bias voltage at a lock-in-frequency of 7.6 kHz. The distance of the two lines is monitored with an OSA.

The measured output of the THz-receiver during the transmission of the generated mm- and THz-waves can be seen in Fig. 6.9(a) for a 200 GHz wave and in Fig. 6.9(b) for a frequency of 1 THz. During the measurement the free space link was intentionally interrupted, represented by the signal drops in Fig. 6.9. As can be seen, the measured power decreases significantly with higher frequencies, due to the conversion efficiency of the used photo mixers. Therefore, a measurement of 2 and 3 THz was not possible. However, the THz-wave itself could not be characterized by means of linewidth and phase noise.

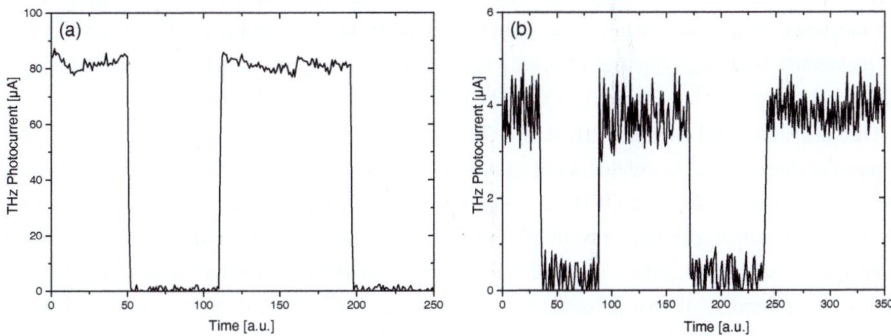

Figure 6.9.: Measured photocurrent at the THz-receiver for 200 GHz (a) and 1 THz (b). At the intervals with zero photocurrent, the free-space link was intentionally interrupted.

The frequency stability of the wave crucially depends on the laser repetition rate and the generated frequency. A change of 1 Hz in the repetition rate of the fiber laser, for example, would modify the frequency of a 1 THz waveform by 13.3 kHz. The frequency drift, caused by temperature, has no influence on the filtering with the DFB laser diodes since the 3 dB bandwidth of SBS is 10–30 MHz and the lasers are locked by the PDH-module. However, the relatively low repetition rate can be carefully stabilized electrically [212], leading to very stable mm- and THz-waves.

In principle, narrow linewidth signals with low phase noise can be easily generated by the electro-optic down-conversion of two lines out of a mode locked fiber laser. The lines have stable amplitude as well as fixed frequency and phase relations between each other. The arbitrary exclusive selection of two lines is enabled by narrow-band SBS amplification, along with the polarization pulling effect. The maximum possible generated frequency just depends on the used photomixer. The possibility of generating high-quality waves, together with the compact light-weight, reliable and stable setup based on standard components of optical telecommunications makes the method especially attractive in the field of high-bitrate wireless communication.

6.2. Nyquist Pulse Generation

With increasing demand for higher data rates in telecommunication networks, the given bandwidth of the communication channels needs to be fully utilized. Within the last decade higher spectral efficiency was obtained by the utilization of higher order modulation formats as well as several multiplexing techniques, accompanied by a drastic increase in the requirements for electrical signal processing. The solution seems to be the combination of several lower rate channels with high spectral efficiency to a single Tb/s superchannel [213], but it requires dense packaging of the sub carriers to fully utilize the given bandwidth.

One possible multi carrier method is orthogonal frequency division multiplexing (OFDM), where the data stream is divided within the available frequency range to a plurality of narrow frequency carriers [214]. For OFDM systems sinc-shaped frequency bands are generated with Fast-Fourier-transform, which leads to rectangular pulses in the time domain. Contrary, a special kind of Nyquist pulses is sinc-shaped in time domain and therefore rectangular shaped in frequency domain. If several rectangular subchannels are combined to a single superchannel, the maximum bandwidth efficiency can be achieved if the spectra of the subchannels are directly adjacent to each other and the bandwidth of the single spectrum is the inverse of the symbol duration. This method for the generation of superchannels is the so called Nyquist wavelength division multiplexing. Recently, the transmission of 32.5 Tbit/s with a spectral efficiency of 6.4 bit/s/Hz, based on Nyquist WDM, have been shown [215]. Under idealized assumptions OFDM and Nyquist WDM have the same sensitivity and spectral efficiency.

However, OFDM requires a much larger receiver bandwidth and proportionally faster speed of the analog-to-digital converters [216]. Furthermore, compared to OFDM, Nyquist WDM tributaries can be transmitted and processed asynchronously, the pulse shaping leads to lower peak-to-average power ratios [217] and is less sensitive to fiber nonlinearities.

Besides high bit rate optical telecommunications many other fields could benefit from sinc pulses. The ideal interpolation function for the perfect restoration of band-limited signals from discrete and noisy data corresponds to sinc-shaped pulses, for instance. Thus, optical sampling devices could be substantially improved [218]. Furthermore, sinc pulses could enable the implementation of ideal rectangular microwave photonics filters [219–221].

Nyquist pulses can be generated electronically, which can be done by an arbitrary waveform generator for instance [222]. However, conventional generators are very limited regarding sampling rate and processor capacity. Another possibility is the optical filtering of a comb generated by a mode locked laser [223], or the generation by optical parametric amplification [224]. Optical Nyquist pulse generation can generate much shorter pulses than electronics. However, most of the reported methods use complex and costly equipment and due to the optical filter characteristics the generated pulses are not sinc-shaped.

According to the theory, a single sinc pulse is unlimited in time domain and therefore non-causal. However, recently it has been shown that under certain conditions, the periodic summation of unlimited sinc pulses leads to a periodic sinc pulse sequence [225]:

$$\sum_{n=-\infty}^{\infty} sinc\left(\pi N \Delta f \left(t - \frac{n}{\Delta f}\right)\right) = \frac{sin(\pi N \Delta f t)}{N sin(\pi \Delta f t)} \quad \circ\!\!-\!\!- \quad \sum_{n=\frac{N-1}{2}}^{\frac{N-1}{2}} \delta(f - n\Delta f) \tag{6.1}$$

with $n = 1, 2, 3, ...$ as an integer number. The Fourier transform ($\circ\!\!-\!\!-$) represents a frequency comb with N equally spaced frequency lines ($\delta(f)$) separated by Δf. Thus, an ideal sinc pulse sequence can be simply obtained by a flat, phase-locked frequency comb with narrow lines. The symbol duration Δt of the sinc pulse is defined by the inverse bandwidth B of the comb $\Delta t = 1/B = 1/(N\Delta f)$ and the repetition rate of the pulses is determined by the frequency spacing between the comb lines $T = N\Delta t = 1/\Delta f$, as can be seen in Fig. 6.10. Such a phase locked frequency comb can be generated via two cascaded Mach-Zehnder modulators (MZM) [225]. The generated sequence of sinc pulses is almost perfectly shaped. However, this scheme can only provide frequency combs with no more than 3 frequency tones using a single MZM and no more than 9 frequency tones using two MZMs in a row. Therefore, the overall bandwidth of such a comb is limited due to the frequency range of the modulators.

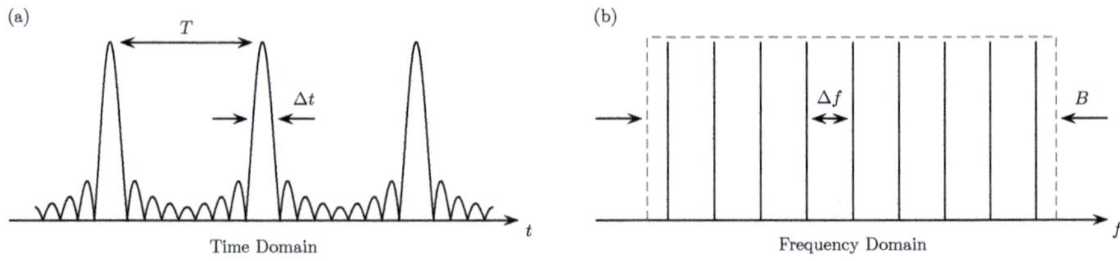

Figure 6.10.: Time and frequency domain representation of a Nyquist pulse train.

The maximum bandwidth can be achieved if the carrier of the first modulator is suppressed. If at least one of the modulators can be driven with a maximum frequency of $f_1 = 40\,\text{GHz}$, the second modulator has to be driven with $f_2 = 2/3f_1$ and the maximum bandwidth is $160\,\text{GHz}$, corresponding to $6\,\text{ps}$ pulses. The pulse duration and the repetition rate of the generated pulses are rather limited. Additionally, the small number of frequency lines result in a Nyquist pulse with a relatively large duty cycle (9.8%), which is unsuitable in applications such as high-speed orthogonal time division multiplexing and arbitrary waveform generation.

The measurement of such a frequency comb and Nyquist pulse train can be seen in Fig. 6.11. The spectral representation is limited by the resolution of the OSA. The width of each comb line depends on the used laser source and is in reality much narrower. The generated Nyquist pulse sequence is almost perfect and fits to the theoretical calculation.

In order to enhance the bandwidth of the frequency comb and the possibilities for the Nyquist pulse generation, multiple carrier frequencies are needed as a source for the two modulators. In principle several DFB laser diodes or an array of other lasers could be used.

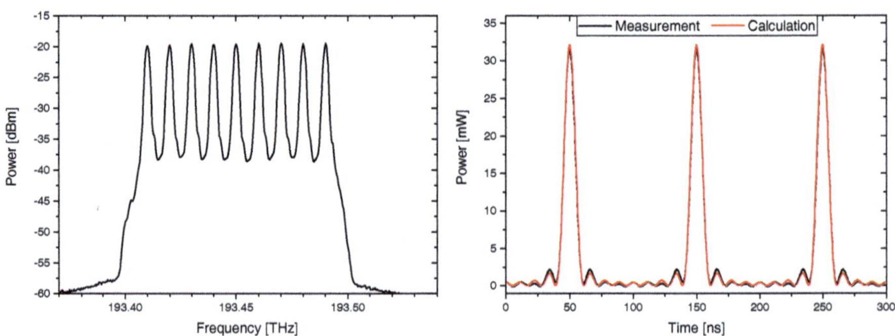

Figure 6.11.: Measurement of a frequency comb and the according Nyquist pulse train generated by coupled MZMs.

116

But, the phases are not locked between the sources and therefore no Nyquist pulses are generated. The ideal source with a large number of equidistant phase locked frequencies is a frequency comb, as can be seen in Fig. 6.12(a) [226]. Therefore, sources like mode locked lasers or electro-optical comb generators could be utilized to generate frequency combs that span several THz. In general the output power distribution of these sources is not equal. Therefore, a number of lines is extracted from the comb and the powers are equalized. The selection of the different lines is again carried out by polarization pulling assisted Brillouin scattering.

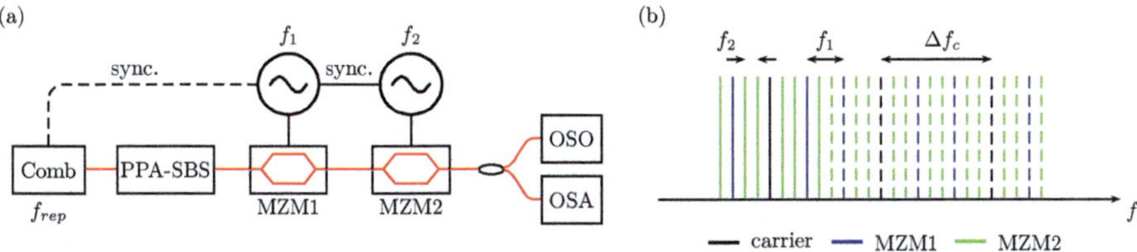

Figure 6.12.: Operation principle for the enhanced Nyquist pulse generation (a) and the generated frequency combs (b) with one carrier (solid) and multiple carriers (dashed).

In order to achieve variable and equal frequency spacing between the comb lines, the separated frequency lines are modulated by the two cascaded MZMs, as can be seen in Fig. 6.12(b). For a mode locked laser the frequency spacing between the comb lines is defined by the repetition rate f_{rep}. Thus, the initial spacing between the comb lines is $\Delta f_c = (n_k + 1) \cdot f_{rep}$, where $n_k = 0, 1, 2, 3, \ldots$ is the number of frequencies between two selected lines. The electrical power of the modulation signal as well as the bias voltage of the first modulator are adjusted to generate two lines around the initial one and to ensure that they have the same power. The modulation frequency for the first modulator f_1 defines the frequency spacing between these lines. Since all lines have to have the same frequency spacing, it follows that $f_1 = 1/3\Delta f_c$. Accordingly, for the frequency of the second modulator it follows that $f_2 = 1/3f_1 = 1/9\Delta f_c$. Thus, for two modulators $f_2 = \Delta f$ and for n modulators the frequency difference between the comb lines is $\Delta f = 1/3^n\Delta f_c$. The parameters of the second modulator are adjusted to generate two equal lines around the initial ones as well. The bandwidth and number of lines of the generated comb is $B = (l - 1) \cdot \Delta f$ and $l = n_l \cdot 3^n$, with n_l as the number of extracted lines from the mode locked laser and n as the number of modulators. For small repetition rates the initial frequency spacing can be made completely tunable for lower frequencies, enabling a broader range in which the comb parameters can be changed. In order to ensure an equal spacing, the number of modulators needs to be increased for narrow frequency spacing if the repetition rate is high, e.g. 10 GHz. But each additional modulator increases the complexity and has insertion losses which decrease the signal to noise ratio. Therefore, the experiment was carried out with just two modulators in a row. The experimental setup can be seen in Fig. 6.13. The MLL during the experiment was again the fs-laser FemtoFErb 1560 with a repetition rate of 100 MHz and an overall bandwidth of 30 THz. A part of the output spectrum is preselected by a WS and subsequently amplified to an output power of 10 dBm. The comb lines pass a polarization controller (PC1) and were coupled into a 10 km AllWave fiber via a circulator (C_2). The individual lines are filtered out by polarization assisted stimulated Brillouin scattering. Thus, for each extracted line a separate pump laser is required. In accordance to the previous setups conventional distributed feedback laser diodes (LD) are used. The pump power and polarization is set by the belonging EDFAs and PCs, respectively. The different pump waves are combined and coupled into the fiber via C_1. The pump lasers are locked to the desired line of the fs-laser by minimum value regulation realized by a software program. The pump wave pulls the SOP of the signal during the SBS process, as described previously. The unamplified components that are not affected by SBS have a different SOP than the amplified ones, and can be separated with the PBS.

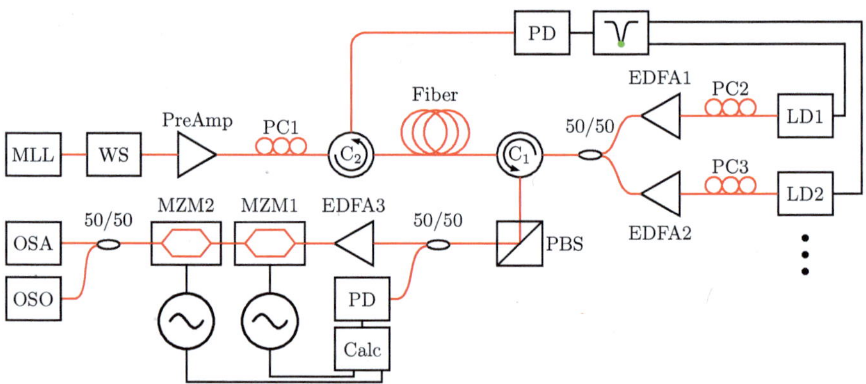

Figure 6.13.: Experimental setup for the flat and flexible frequency comb generation out of a MLL.

The extracted frequencies out of the comb are amplified by EDFA3 and processed by the two MZMs. Thereby, MZM1 was a 40 Gb/s and MZM2 a 10 Gb/s intensity modulator. As described above, the two modulators are driven with 1/3 and 1/9 of the frequency spacing between the extracted comb lines. Therefore, the frequency difference between the two extracted lines is measured with a PD followed by an ESA. The RF signals for the two modulators are calculated and adjusted accordingly (Calc). Finally, the frequency comb is measured with an OSA and the time domain representation is analyzed with an optical sampling oscilloscope (OSO). At the beginning, two lines were extracted out of the fs-laser spectrum with a frequency spacing of 90 GHz. Therefore the modulation frequencies are calculated to 30 GHz for MZM1 and 10 GHz for MZM2. In order to generate just one sideband at each side of the extracted line, the RF power was 12 dBm ($0.5\,V\pi_{RF}$) for MZM1 and 13 dBm ($0.55\,V\pi_{RF}$) for MZM2. The bias voltage of the modulator was adjusted to align all lines to the same power level and was 1.09 V ($0.33\,V\pi_{DC}$) for MZM1 and 1.51 V ($0.3\,V\pi_{DC}$) for MZM2. This resulted in a frequency comb with 18 lines with a spacing of 10 GHz and an overall bandwidth of 170 GHz, as can be seen in Fig. 6.14(a). The power fluctuations of the different frequencies in the comb appear to be below 0.6 dB. The time domain representation

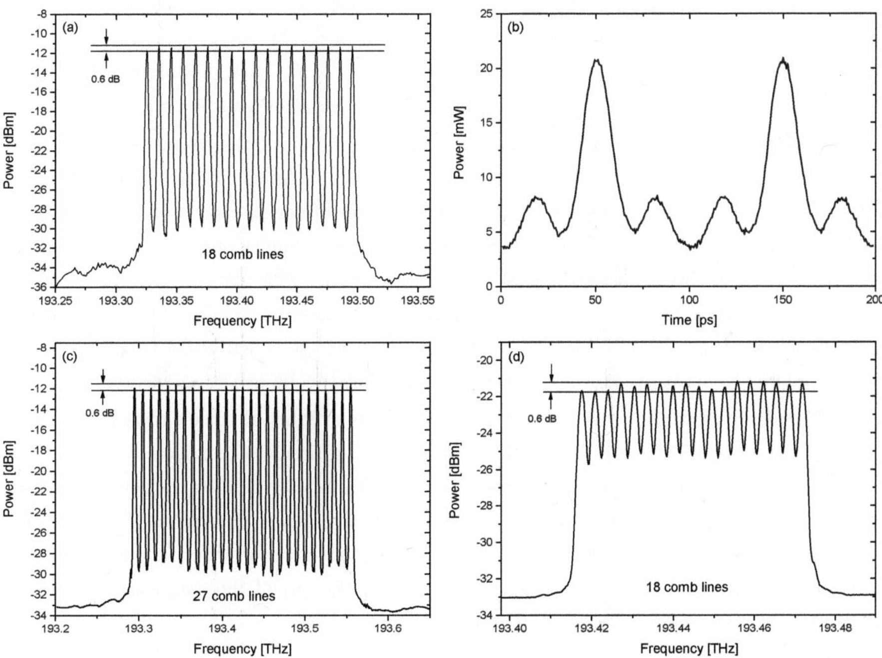

Figure 6.14.: Generated frequency combs with (a) 18 lines and 10 GHz spacing, (c) 27 lines with 10 GHz spacing and (d) 3 GHz spacing. The time domain representation of (a) can be seen in (b).

of the generated comb can be seen in Fig.6.14(b).

Furthermore, three lines out of the comb generator spectrum were extracted. To achieve an equal spacing between the three extracted lines, the difference frequencies are monitored. The selection of the third line is adjusted until just one difference frequency of 90 GHz appeared after the PD. Once again, the frequencies for the two modulators are calculated to be 30 GHz and 10 GHz. The RF power remains the same and the bias voltage of the modulators is adjusted again to achieve a flat frequency comb. The result is a frequency comb with 27 lines and a frequency spacing of 10 GHz, as can be seen in Fig. 6.14(c). The overall bandwidth of the comb is 260 GHz and the power variations of the several comb lines are within 0.6 dB.

In order to prove the arbitrarily generation of frequency combs the initial spacing between the extracted modes Δf_c is changed to 27 GHz. Therefore the frequencies for the modulators are calculated to be 9 GHz for MZM1 and 3 GHz for MZM2. For two extracted lines out of the fs-laser this leads to a flat frequency comb with 18 lines, as shown in Fig. 6.14(d). The power variations are as low as 0.6 dB. The long term stability of the amplitudes of the comb lines is within 3 dB over several hours. However, the drift of the operating point due to the DC bias voltage can be compensated with a bias controller for each modulator. Therefore, high long term stability in the range of < 0.5 dB power deviation can be realized.

The time domain representation, measured with the OSO, of the frequency comb with 18 lines and a spacing of 10 GHz is shown in Fig. 6.14(b). As can be seen, some kind of pulse train is generated, originated in the locked phases of the modulation. But, they are not locked to the fs-laser. In order to generate sinc-shaped Nyquist pulse sequences, the RF generators driving the modulators need to be synchronized with the repetition rate of the fs-laser. Possibly this can be realized with a phase locked loop or a simple frequency divider to adapt the measured repetition rate to the frequency of the synchronization input of the radio frequency generators. However, these concepts need to be investigated further.

For the sake of simplicity, the fs-laser was replaced by a Fabry-Perot electro-optical comb generator driven by a DFB laser diode. The comb generator needs to be driven by an electrical signal that corresponds to his resonance, which simultaneously defines the repetition rate. During the measurement it was set to f_{rep}=9.52 GHz with a power of 10 dBm. This electrical generator was then synchronized with the other ones, that generate the frequencies for the coupled modulators. Therefore, the extracted frequencies out of the comb are completely phase locked to the modulation signals of the cascaded MZMs.

As previously, the measurement started with the extraction of two line out of the comb generator. The extracted lines had a spacing of $\Delta f_c = 57.12$ GHz. This results in modulation frequencies of $f_1 = 19.04$ GHz and $f_2 = 6.346$ GHz for the two modulators. The first modulator was driven with a RF power of 5 dBm and a bias voltage of 1.11 V and the second one with

13 dBm and a bias voltage of 2.08 V. The generated frequency comb with a flatness of 0.6 dB and a SNR of 36 dB can be seen in Fig. 6.15(a). The according time domain signal, basically the Nyquist pulse sequence, can be seen in Fig. 6.15(b). As can be seen, the comb consists of 18 lines which correspond to 17 zero-crossings of the pulse sequence. The overall bandwidth of the comb is 114.228 GHz, resulting in a pulse width of $1/N\Delta f = 8.75$ ps. The repetition rate of the pulses is $T_R = 1/\Delta f = 157.6$ ps. Accordingly, the duty cycle is 5.5%. The red dashed lines in all parts of Fig. 6.15 show the calculation of a perfect shaped Nyquist pulse sequence. As can be seen, the generated pulse sequence fits very well with the theoretical prediction.

Furthermore the number of extracted frequencies out of the comb generator was increased. The results for the extraction of three lines can be seen in Fig. 6.15(c) and (d). The repetition rate as well as the frequencies for the modulators remains the same. This results in a flat frequency comb with 27 lines with a bandwidth of 171.342 GHz and a spacing of 6.346 GHz. The flatness of the comb is 1.04 dB and the SNR is 27.4 dB. The pulse width is 5.8 ps and the repetition rate is again 157.6 ps, which leads to a duty cycle of 3.6%. It must be noted that there are some small distortions in the generated pulse-sequence. The zero-crossing are at the correct position, but they do not reach the zero-level at every designated point. This can be addressed to the not equal power distribution of the comb.

The generated signals for 5 extracted lines can be seen in Fig. 6.15(e) and (f). The comb consists of 45 lines with a spacing of 6.346 GHz. This results in a total bandwidth of 285.57 GHz. The repetition rate is again 157.6 ps. Due to the large bandwidth of the generated comb, the pulse width is further decreased down to 3.5 ps. In this case the duty cycle is 2.2%. The measurement fits well with the calculations. The small deviations are again a result of the uneven power distribution of the initial lines.

In principle, the repetition rate of the pulses as well as the pulse width can be randomly changed. The theoretical limits are the spectral width of the MLL or the optical bandwidth and nonlinearity of the used modulators. The generated sinc pulse sequences have such a high quality that they can bring benefits to optical communications with the maximum baud rate. First experiments have shown the straight line transmission of five 112.5 Gbd Nyquist OTDM dual polarization quadrature phase-shift keying channels over a dispersion uncompensated link up to 640 km [227]. The utilization of the Nyquist pulses effectively improves the spectral confinement of the modulated OTDM signal, providing a minimum intercarrier crosstalk penalty of 1.5 dB in baud-rate-spaced Nyquist-WDM systems. Additionally, the sinc-shaped Nyquist pulse trains are well suited for all-optical buffering, microwave photonic filters, arbitrary waveform generation and many other fields.

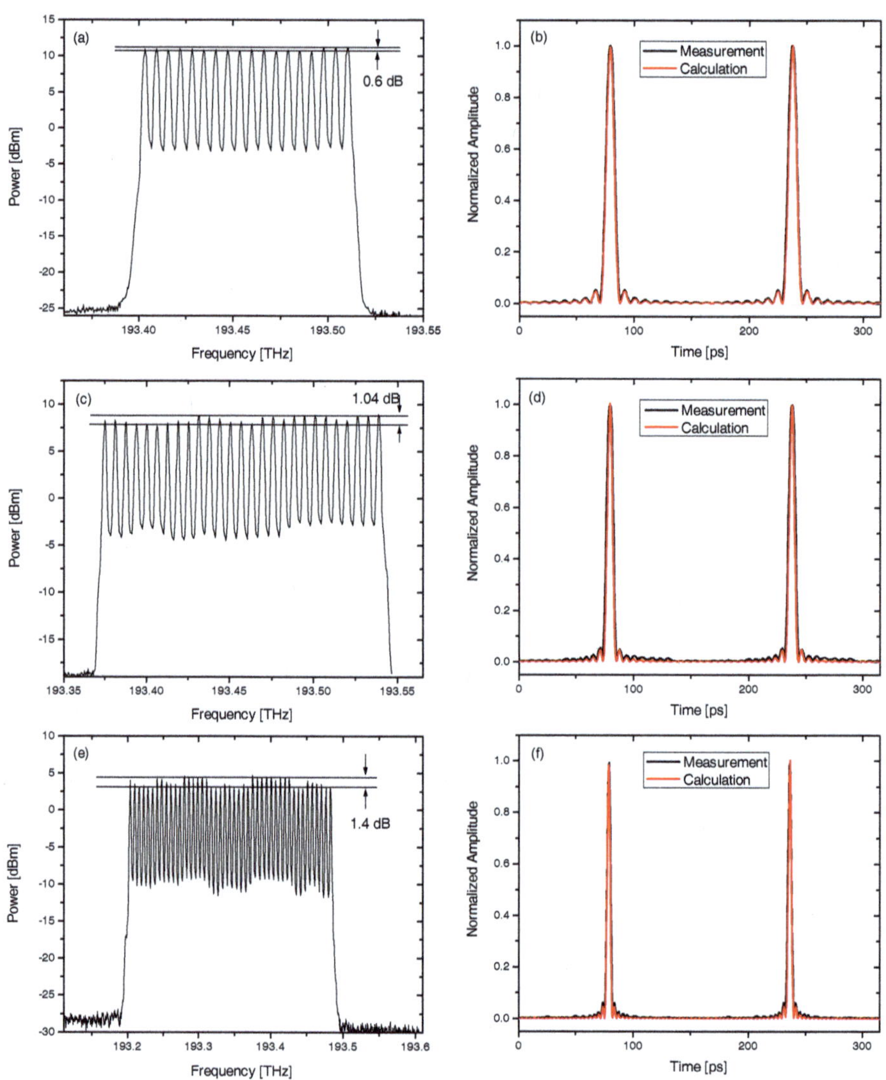

Figure 6.15.: Generated frequency combs with a spacing of 6.346 GHz and the corresponding Nyquist pulse trains for different numbers of extracted lines out of the comb generator.

7

Summary and Outlook

In summary, this thesis has introduced several schemes for the bandwidth reduction of stimulated Brillouin scattering, as well as, their utilization in different applications. The first approach for the bandwidth reduction was realized by a multi stage SBS system, where the bandwidth could be decreased down to 5.8 MHz within 3 stages. Second, the saturation characteristics of Brillouin scattering were exploited for the bandwidth reduction, where a narrow band frequency domain aperture covers the gain. This method was capable to reduce the bandwidth down to approximately 3 MHz, which equals only 15% of the original bandwidth. Within the third method, the Brillouin bandwidth was reduced by the superposition of the gain with two losses. Again, the bandwidth was decreased significantly down to 3.4 MHz. All presented methods were realized in standard optical fibers at room temperature with off-the-shelf equipment, emphasizing the simplicity. These narrow bandwidths were exploited in several applications and enhanced their performance significantly.

The first application was focused on optical spectrum analysis, the key technique for characterization of optical devices and systems. The utilization of normal SBS enabled a resolution of 10 MHz, which is 200 times better than conventional grating based analyzers. With the help of the investigated methods for the Brillouin bandwidth reduction the resolution could be enhanced even further, leading to Brillouin optical spectrum analyzers with a resolution of 3 MHz. Additionally, the polarization pulling attributes of Brillouin scattering were applied in order to form sharper filter characteristics and enhanced the dynamic range of a classic Brillouin based spectrum analyzer by more than 30 dB. Pushing the limits even further, the narrow bandwidth filtering of polarization assisted Brillouin scattering was combined with heterodyne detection. First experiments with a narrow linewidth fiber laser as local oscillator have confirmed the potential of this method, demonstrating a resolution of 1 kHz and revealing almost every spectral detail of different signals under test. Compared to the classical grating based spectrum analyzer, with this approach the resolution was enhanced by six orders of magnitude or 2 million times, respectively. The achieved spectral resolution is higher than in all conventionally available devices. Finally, this measurement procedure was highlighted in „Advances in Engineering" in February 2016.

The second presented application, named Quasi-Light-Storage, was a newly introduced technique for the tunable delay and storage of optical data packets. Within this simple method, the SBS was utilized as frequency selective amplifier, where the bandwidth inversely defines the maximum storage time. For the normal Brillouin gain this led to maximum storage times of 100 ns. The storage of several 8 bit data packets with a data rate of 1 Gbps was demonstrated, showing rather low distortions of the delayed packets. Emphasizing the significance of this approach, it was selected by „Nature Photonics" as a „Research Highlight" in October 2009. It has been shown, that the proposed methods for the Brillouin bandwidth reduction could enhance the storage time significantly. The QLS expanded to a multi stage system reached a maximum storage time of 160 ns, whereas the utilization of a superimposed gain with two losses led to an increased storage time of 140 ns. The distortions were still very low and almost no pulse broadening occurred. A further increase of the storage time up to several hundred nanoseconds was achieved by a feedback system. Furthermore, the storage of phase modulated data packets was shown, reaching storage times up to 60 ns. Again, the significance of this enhancement was mentioned in „Nature Photonics" in October 2012 as „Research Highlight".

Within the last investigated application, SBS was exploited as narrow bandwidth optical filter for the processing of optical frequency combs. The filtering of two lines out of the comb led to the tunable generation of high quality mm-waves up to 110 GHz with a very low phase noise of -104 dBc/Hz in the initial measurements. Further experiments have shown that the phase noise can be as low as -134 dBc/Hz [228, 229]. Furthermore, this method was adapted to the THz range, enabling high quality carrier waves for ultra high data rate wireless communications. First transmission experiments were demonstrated at a carrier frequency of 1 THz. Further processing of the extracted comb lines was done by two coupled Mach-Zehnder modulators, leading to the generation of ideal rectangular and flat frequency combs. Within the experiments ideal combs consisting of up to 45 lines with variable spacing and power fluctuations below 0.5 dB were achieved. Afterwards the comb source was synchronized with the electrical generator that provided the signal for the modulators, leading to phase locked frequencies in the generated comb. The time domain representation revealed almost ideal sinc-shaped Nyquist pulse trains with pulse widths down to 3.5 ps and a minimum duty cycle of 2.2%. The main advantage of the sinc-shaped pulses for optical communication is the perfect rectangular spectrum and the zero intersymbol interference. Within first experiments the straight line transmission of five 112.5 Gbd Nyquist OTDM dual polarization quadrature phase-shift keying channels was demonstrated over a dispersion uncompensated link up to 640 km [227].

Outlook

This thesis has investigated the bandwidth reduction of stimulated Brillouin scattering and showed several applications. During the measurements different problems were found and concepts for future improvement have been developed. These recommendations, perspectives and possible new applications are summarized here.

The bandwidth of stimulated Brillouin scattering could be reduced successfully down to 3 MHz with an optical aperture. Therefore, the signal needs to be overlaid by this strong saturation signal. During detection the aperture needs to be removed from the signal, which is realized by lock-in detection so far. For applications that only employ the narrowband amplification, e.g. the optical spectrum analyzer, this circumstance does not matter. But for signal processing applications, e.g. QLS, this leads to significant distortions. For future applications, the involved aperture needs to be deleted completely before or during detection. This could be realized by different schemes like phase modulation or balanced detection.

The presented approach for optical spectrum analysis enhanced the resolution significantly. Currently, a costly fiber laser with a narrow linewidth and a rather limited tuning range is utilized. This circumstance can be avoided in the future by utilizing a Brillouin fiber laser instead. Therefore, a conventional, low cost and tunable DFB laser diode will act as pump source for the prefiltering of the signal, as well as pump source for a Brillouin fiber laser in an additional short fiber. These lasers achieve linewidths below 1 kHz as well. For a continuous tuning over a broad range an array of DFB laser diodes could be used.

The QLS have shown the tunable storage of amplitude and phase modulated signals. In order to implement this approach in future communication systems, the applicability to higher order modulation formats like QAM, that employ phase and amplitude information at the same time, needs to be investigated further. Theoretically, the QLS can be applied for these formats as well without any problems.

First transmission experiments at a carrier frequency of 1 THz were demonstrated over a distance of 24 cm. In principle, the performance of the system is mainly restricted by the used THz mixers. They have a very limited output power, a high beam divergence and especially the receiver is not equipped with a transimpedance amplifier. Future investigations need to utilize transmitters with higher powers, like uni-traveling carrier photo diodes, as well as, different receiver structures, or employ a high bandwidth transimpedance amplifier to the existing THz-receiver, to increase the possible bandwidth for high data rate wireless communication experiments. It is expected that the quality of the THz-waves is also exceptional and could be also used for phase noise sensitive higher order modulation formats, enabling wireless multi Tbps transmissions.

The generation of almost ideal sinc-shaped Nyquist pulse sequences and the utilization in WDM transmission systems showed very promising results. Future measurements will include higher data rates, probably more channels and the integration on a silicon on insulator platform. Besides high bit rate optical telecommunications, this technique for generating almost ideal sinc-shaped Nyquist pulses could be exploited for optical sampling, which is the first step to convert an analogue optical signal into a digital electrical signal. Thereby, the unknown input signal would be processed by the coupled modulators, resulting in a multiplication of the unknown signal with the Nyquist pulse train. Afterwards the peak powers of the separate pulses will be analyzed. In principle, all sampling parameters could be simply and fast tuned in the electrical domain and a parallelization of sampling in time and frequency domain is straight forward. Contrary, the multiplexing of several time delayed Nyquist pulse sequences could lead to an all optical arbitrary waveform generator. With the proposed methods all parameters like repetition rate, duty cycle and pulse width could be easily adapted.

In general, stimulated Brillouin scattering and its applications will have a bright future. Recently, it was shown that SBS is realizable on silicon chips [230] as well as on chalcogenide based chips [231]. This will enable the integration of existing Brillouin based applications and possibly enable even new applications. First proof of concept measurements with this new technology were shown in the field of Brillouin based microwave photonic filters [232], the traditional Slow- and Fast-Light [233], microwave signal processing and generation [234] as well as the realization of Brillouin lasers on a chip [235]. Theoretical, this technology can lead to a Brillouin based optical spectrum analyzer with the size of a matchbox, including all components for measurement and data acquisition, or a chip that includes multiple structures for the tunable storage of optical data packets. Last but not least, all the introduced methods for the Brillouin gain bandwidth reduction should be realizable with the chip technology. However, further investigations in this regard are necessary.

A. Convolution Theorem

The convolution is a calculation method, related to the multiplication, for two functions that can be used both in the time domain and the frequency domain. The convolution integral or convolution product is described by [236]:

$$y(t) = x_1(t) * x_2(t) = \int\limits_{-\infty}^{+\infty} x_1(\tau)x_2(t - \tau)d\tau \tag{1}$$

A convolution in the time domain corresponds to a multiplication in the frequency domain:

$$s(t) * h(t) \quad \circ\!\!-\!\!\bullet \quad S(f) \cdot H(f) \tag{2}$$

and a convolution in the frequency domain corresponds to a multiplication in the time domain:

$$s(t) \cdot h(t) \quad \circ\!\!-\!\!\bullet \quad S(f) * H(f). \tag{3}$$

The δ-function, which is in honor of the mathematician Paul A. M. Dirac also called Dirac-function, is in the mathematical sense actually no function, but a distribution [236]. This means that a function value is not obtained by insertion of an argument, but by performing an arithmetic rule. There are several ways to define the delta function, a variant is:

$$\delta(t) = 0 \quad \text{for} \quad t \neq 0$$
$$\int\limits_{-\infty}^{\infty} \delta(t)dt = 1 \tag{4}$$

The Dirac function can be considered as a rectangular function with the area 1. The rectangular function has the level 1 and the width T_0. Multiplying the function with $1/T_0$

and then leave T_0 approaching zero, the result is a square pulse $\delta(t)$ which is infinitely high and infinitely thin:

$$\lim_{T_0 \to 0} \frac{1}{T_0} rect\left(\frac{t}{T_0}\right) = \delta(t) \tag{5}$$

According to Eqn. 4, the Dirac pulse is defined at the point $t = 0$. During the multiplication with an arbitrary function $x(t)$, the Dirac pulse defines a function value at the point $t = 0$. All parts of $x(t)$ for $t \neq 0$ are multiplied by zero and are therefore hidden. This behavior is referred to as the sampling characteristics of the Dirac pulse.

If the time variable of the Dirac pulse is weighted by a factor of b, the result is a stretched Dirac pulse $\delta(bt)$. To derive the properties of the stretched Dirac pulse, a convolution with an arbitrary signal $s(t)$ is formed. For the convolution with an arbitrary signal $s(t)$ it follows:

$$\delta(bt) * s(t) = \int_{-\infty}^{\infty} \delta(b\tau)s(t - \tau)d\tau = \frac{1}{b}s(t). \tag{6}$$

Accordingly, the result of the convolution is also weighted. The width of the folded signal is reduced in the frequency range to $\frac{1}{b}$. The other way round, a weighted Dirac function in the frequency domain would lead to a limited time domain response with the envelope $\frac{1}{b}$.

With the help of the Dirac pulse, it is possible to form the so-called sampling or Dirac pulse sequence. Therefore, multiple duplicates shifted by T are added to the original Dirac pulse. Accordingly, a Dirac pulse sequence $\delta_T(t)$ consists of an infinite number of equidistant arranged Dirac pulses. The definition of the Dirac pulse sequence is [237]:

$$\delta_T(t) = \sum_{n=-\infty}^{\infty} \delta(t - nT), \tag{7}$$

where T indicates the distance between the individual Dirac pulses. An example for a Dirac pulse sequence can be seen in Fig. A.1. It should be noted that the Dirac pulse sequence has, according to Eqn. 7, a zero phase angle equal to zero. Thus, the Dirac pulse occurs for the index $n = 0$ at the point $t = 0$. Commonly, the Dirac pulse sequence is also called Dirac comb.

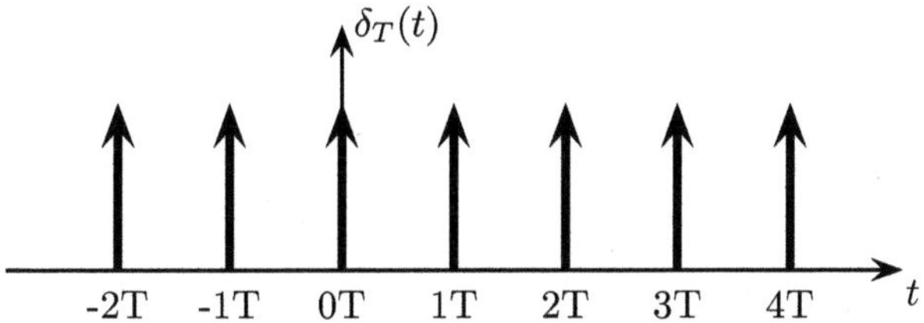

Figure A.1.: Dirac pulse sequence

The convolution of a non-periodic function $f(t)$ with the Dirac pulse sequence $\delta_T(t)$ is given by the expression:

$$f_T(t) = \delta_T(t) * f(t) = \int\limits_{-\infty}^{+\infty} \delta_T(\tau) \cdot f(t - \tau) d\tau \tag{8}$$

Substituting Eqn. 7 in Eqn. 8 results in:

$$\begin{aligned} f_T(t) =\delta_T(t) * f(t) &= \int\limits_{-\infty}^{+\infty} \sum_{n=-\infty}^{\infty} \delta(\tau - nT) \cdot f(t - \tau) d\tau \\ &= \sum_{n=-\infty}^{\infty} \int\limits_{-\infty}^{+\infty} \delta(\tau - nT) \cdot f(t - \tau) d\tau = \sum_{n=-\infty}^{\infty} f(t - nT). \end{aligned} \tag{9}$$

The convolution integral is divided into an infinite sum of convolution integrals. For each index n of the sum, a convolution integral with a correspondingly shifted Dirac pulse needs to be calculated. For that the selective property of the Dirac pulse can be exploited. The Dirac pulse samples under the integral for $\tau = nT$ the value $f(t - nT)$.

The convolution of a non-periodic function $f(t)$ with a Dirac comb $\delta_T(t)$ leads to a copying of the function $f(t)$ with the period T, as illustrated in Fig. A.2. If the function $f(t)$ differ only in a finite time interval from zero, and this interval is shorter than T, a periodic extension of $f(t)$ will be produced. Otherwise, the periodically shifted functions $f(t - nT)$ will overlap. The index T in $f_T(t)$ indicates a periodic function, as for the Dirac comb.

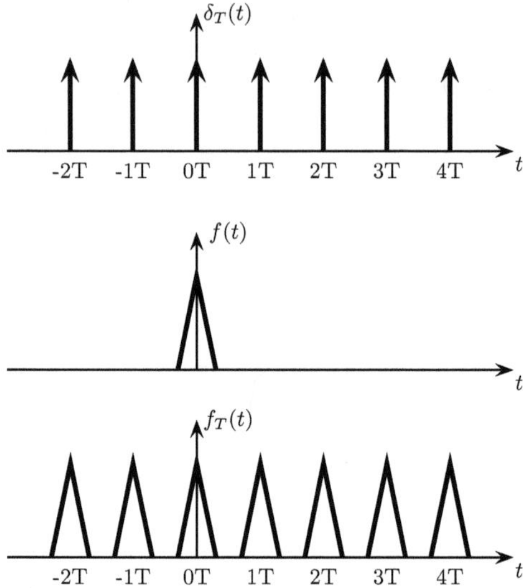

Figure A.2.: Convolution of a non-periodic function $f(t)$ with a Dirac comb $\delta_T(t)$.

The multiplication of a non-periodic function $g(t)$ with a Dirac comb $\delta_T(t)$ can be described by [237]:

$$g_a(t) = \delta_T(t) \cdot g(t) = \sum_{n=-\infty}^{\infty} \delta(t - nT) \cdot g(t) = \sum_{n=-\infty}^{\infty} \delta(t - nT) \cdot g(nT). \qquad (10)$$

During the multiplication with a Dirac comb, the function $g(t)$ can be replaced with the function value at the point t, if the argument of the Dirac comb is not zero, namely the point $t = nT$. Figure A.3 shows the Dirac comb $\delta_T(t)$, a non-periodic function $g(t)$ and the product of both. The result of the multiplication is a series of Dirac pulses at the points nT with the weights $g(nT)$, where n assumes all integer values. The Dirac pulses of $g_T(t)$ are at the same positions as for the function $\delta_T(t)$. The function $g(t)$ forms an envelope over the differently weighted Dirac pulses. Generally, this multiplication with a Dirac comb is called ideal sampling.

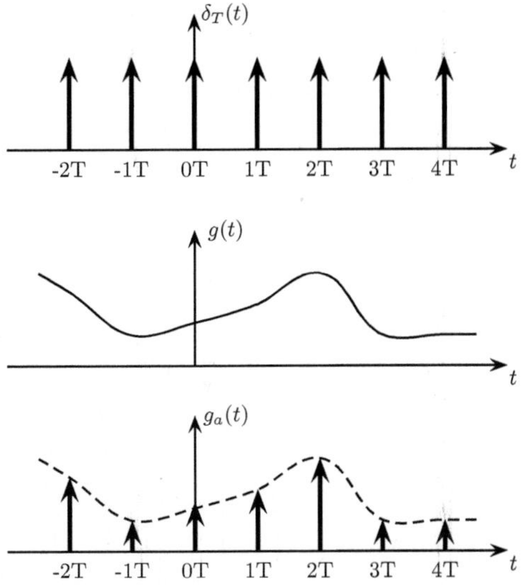

Figure A.3.: Multiplication of a function $g(t)$ with a Dirac comb $\delta_T(t)$.

B. QLS Frequency Comb Generation

The quality of the frequency comb plays an essential role for Quasi-Light Storage. Particularly the flatness of the comb, as well as the number and spacing of the individual lines in the frequency comb are of interest. An uneven power distribution of the comb lines lead to false sampling and therefore to significant distortions in the generated copies. For distortion-free QLS the width of the frequency comb must cover the entire spectral range of the signal and the number of frequencies in the comb must at least equal the number of bits in the data packet. Accordingly, a frequency comb with a width of 1 GHz and at least 8 lines is required for a data packet with 8 bit at a data rate of 1 Gbps.

For the sake of simplicity, the generation of the frequency comb should utilize just a single modulator. Additionally, the quality of the generated comb needs to be as high as possible. During the investigation of QLS several methods for the generation of frequency combs were developed and tested. The first and easiest way for the comb generation is the generation of multiple sidebands through electro-optical modulation. By a sufficiently high power of the electric signal several sidebands are generated, which results in a frequency comb. The power distribution of the spectral lines is determined by the Bessel functions. Therefore the power decreases for higher order sidebands. Depending on the setting of the bias voltage of the modulator, the ratios of the various lines can be changed in the comb. Additionally, using a dual drive modulator and phase shifted driving of both arms, the frequency comb can be further flattened [190]. The corresponding setup and the generated frequency comb are shown in Fig. B.1. The optical carrier is provided by a DFB laser diode. The input polarization for the modulator is adjusted by a polarization controller (PC). The generated comb is analyzed with heterodyne detection and an ESA. As can be seen, there are significant differences in the power distribution of the comb lines. This has direct impact on the quality of the copies produced by the QLS. With this method of frequency comb generation first results for the QLS were achieved [160]. However, this comb leads to obvious distortions.

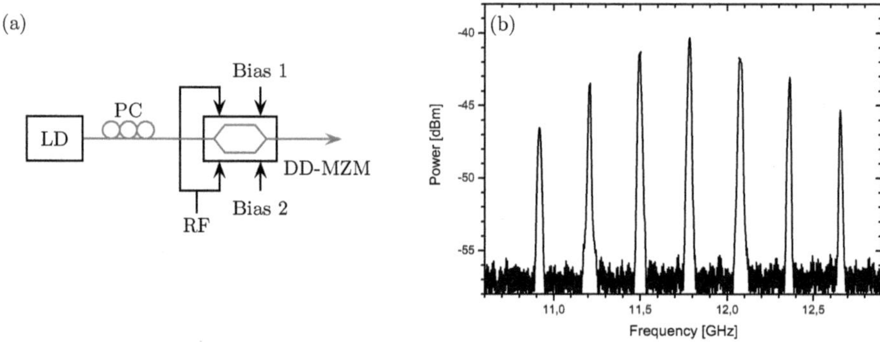

Figure B.1.: Frequency comb generation with a dual drive Mach-Zehnder modulator.

For a better quality frequency comb the electrical input signal for the modulator needs to be adapted. In principle, a multiple number of input frequencies can be generated by electrical frequency multiplication. The fundamental frequency determines the distance between the individual lines in the comb. By any number of mixers, multipliers and adders a frequency comb of any width can be produced. The maximum width of the comb is limited by the working range of the used components. However, for the generation of large combs a large number of electrical components is necessary and increases the complexity significantly. For the conversion into the optical domain with a MZM, the signal must be amplified. This leads to several mixing products due to the nonlinear response of the amplifier. The generated frequency comb is illustrated in Fig. B.2. As can be seen, the power differences of the individual frequency components amounts to at least 5 dB. In addition, the comb is not rectangular shaped due to the mixing products from the electrical amplifier.

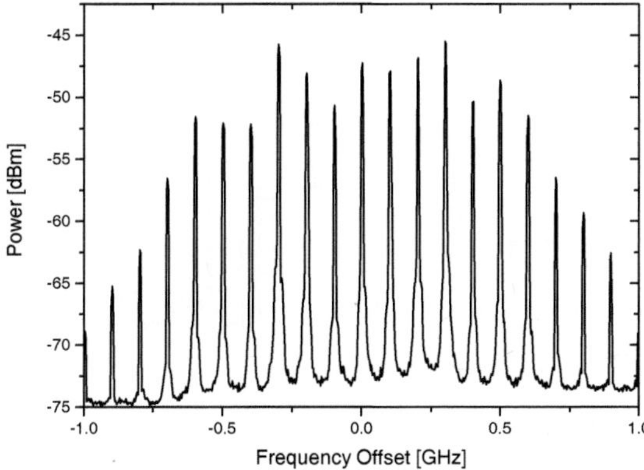

Figure B.2.: Frequency comb generated by electrical frequency mixing and a single electro-optical modulator.

Reduced complexity can be achieved by the utilization of an opto-electronic oscillator (OEO) [238], which is capable of generating repetitive electric sine waves and repetitive optical frequencies with a fixed distance, respectively. In principle, the OEO utilizes the transmission characteristics of a modulator in combination with a fiber optic delay line to convert the continuous light energy of a laser into an electrical frequency or high-frequency signals. Therefore, the output signal of a laser diode is coupled into an MZM, as illustrated in Fig. B.3(a). The signal is then passed through a correspondingly long fiber and received by a photodetector (PD). The PD has a limited bandwidth and acts as a low pass. Alternatively, additional electrical filters can be used to suppress unwanted frequency components. The output of the photodetector is amplified and forwarded as electrical input signal to the modulator. This leads to permanent oscillations with a frequency dependent on the length of the delay line, the bias settings of the modulator and the filter characteristics of the PD. The generated frequency comb can be seen in Fig. B.3(b). The bandwidth of the photodetector was $\approx 1\,\mathrm{GHz}$ and the length of the delay lines was $3\,\mathrm{m}$. All connections between the electrical equipment were kept as short as possible. For optimization an electrical phase adjuster is added. The spacing of the individual frequency components within the comb is $66\,\mathrm{MHz}$. In the central region, the power deviation of the different lines is only 1-2 dB. Due to the filter characteristic of the photodetector, the edges of the frequency comb decline slowly at the sides. Thus, no uniform sampling is possible. Additionally, the tuning of the comb is rather complicated since it depends on the fiber length that is used for the delay.

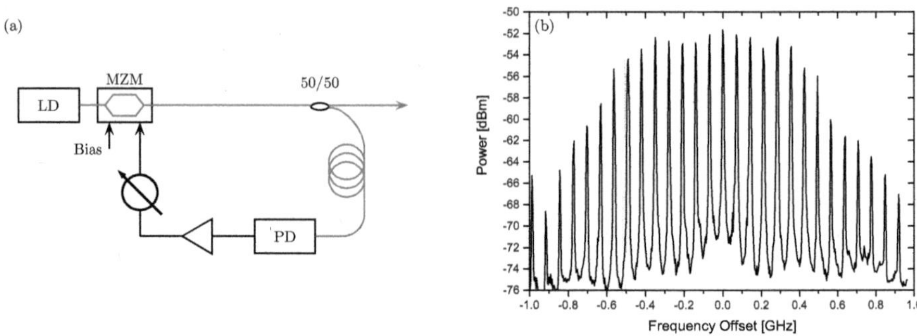

Figure B.3.: (a) experimental setup of an opto-electric oscillator and (b) the generated frequency comb.

All previously described methods for frequency comb generation have significant disadvantages for the QLS. Either the comb does not have enough lines or the individual lines have large deviation in the power distribution. Additionally, these combs are not rectangular shaped, as required for many applications. Far more ideal frequency combs can be easily generated by the utilization of an arbitrary waveform generator (AWG). Thereby, the frequency spacing of the individual lines in the comb as well as the number of lines in the comb can be set arbitrarily. The use of an AWG enables direct generation of multitone signals (multi-frequency signals), which represent a composite of a plurality of sine waves with different frequencies. The basic representation of a multitone signal in time and frequency domain can be seen in Fig. B.4. The output time domain signal of the AWG is represented

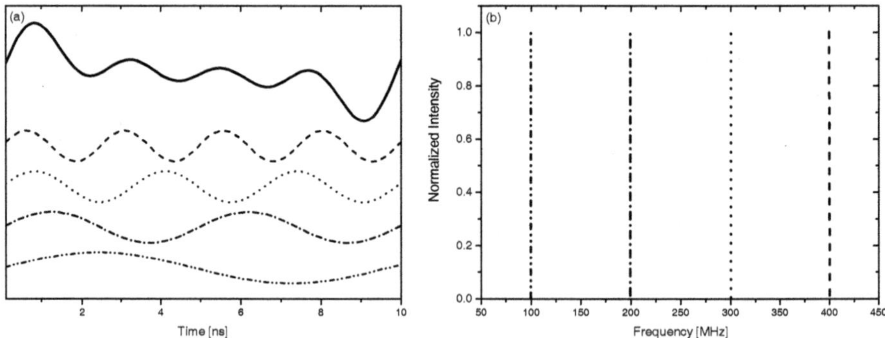

Figure B.4.: Formation of the multitone signal in the time and frequency domain.

by the thick green curve, which is an summation of the sine waves with different frequencies.

In principle, the absolute value of this signal forms a periodic sinc-function. In normal operation mode the AWG continuously repeats the output signal. Accordingly, a periodic sinc-function is sufficient to generate the frequency comb, as described in section 6.2. Thereby the pulse width defines the total spectral width of the comb and the repetition rate defines the frequency distances of the lines. The transformation into the optical domain is again carried out with a single modulator, where a number of sidebands according to the electrical signal are generated on both sides of the optical carrier, e.g. the 4 lines from Fig. B.4 result in an optical frequency comb with 9 lines. The bias voltage of the modulator is adjusted to decrease the carrier in order to generate a flat frequency comb. The generated frequency combs for the usage with the QLS are shown in Fig. B.5. The frequency spacing for both combs is 100 MHz. The multitone signal for (a) consists of 4 frequencies and for (b) of 6 frequencies, which results in a total bandwidth of the comb in the optical domain of (a) 900 MHz and (b) 1300 MHz. As can be seen, the power deviation between the lines is <0.8 dB. The main advantage of this method is the perfect rectangular shaped frequency comb. There are no spectral components on the sides which might distort the signal. The quality of the QLS could be significantly increased. Therefore, in the majority of the experiments this frequency comb was used.

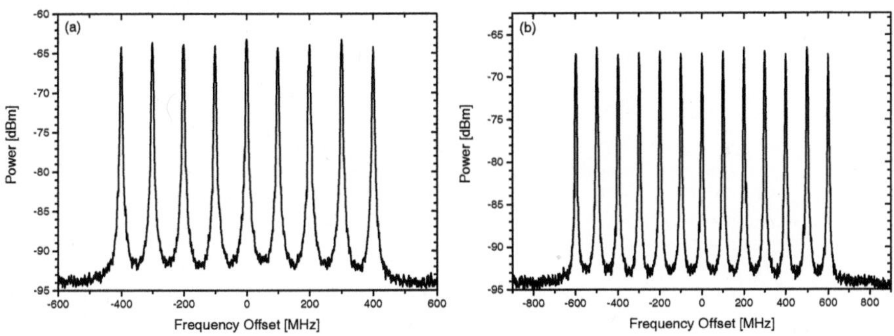

Figure B.5.: Generated frequency comb with a multitone signal provided by an arbitrary waveform generator.

References

[1] G. Agrawal, *Nonlinear Fiber Optics (Optics and Photonics)*, Academic Press, San Diego, 2001.

[2] B. E. A. Saleh and M. C. Teich, *Grundlagen der Photonik*, Wiley-VCH, Berlin, 2 ed., 2008.

[3] S. P. Singh and N. Singh, "Nonlinear effects in optical fibers: Origin, management and applications," *Prog. Electromagn. Res.*, 73, 249–275, 2007.

[4] R. W. Boyd, *Nonlinear Optics, Third Edition*, Academic Press, 3 ed., 2008.

[5] R. Engelbrecht, *Nichtlineare Faseroptik*, Springer, 2014.

[6] T. Zhu, X. Bao, L. Chen, et al., "Experimental study on stimulated Rayleigh scattering in optical fibers," *Opt. Express*, 18(22), 22958–22963, 2010.

[7] L. Dong, "Stimulated thermal Rayleigh scattering in optical fibers," *Opt. Express*, 21(3), 2642–2656, 2013.

[8] O. Frazão, C. Correia, M. R. Giraldi, et al., "Stimulated Raman scattering and its applications in optical communications and optical sensors," *Open Optics Journal*, 3, 1–11, 2009.

[9] S. H. Baek and W. B. Roh, "Single-mode Raman fiber laser based on a multimode fiber," *Opt. Lett.*, 29(2), 153–155, 2004.

[10] H. Rong, R. Jones, A. Liu, et al., "A continuous-wave Raman silicon laser," *Nature*, 433(7027), 725–728, 2005.

[11] H. Rong, A. Liu, R. Jones, et al., "An all-silicon Raman laser," *Nature*, 433(7023), 292–294, 2005.

[12] P. Hansen, L. Eskildsen, S. Grubb, et al., "Capacity upgrades of transmission systems by Raman amplification," *IEEE Photon. Technol. Lett.*, 9(2), 262–264, 1997.

[13] Y. Aoki, "Properties of fiber Raman amplifiers and their applicability to digital optical communication systems," *J. Lightw. Technol.*, 6(7), 1225–1239, 1988.

[14] S. P. Singh, R. Gangwar, and N. Singh, "Nonlinear scattering effects in optical fibers," *Prog. Electromagn. Res.*, 74, 379–405, 2007.

[15] L. Brillouin, "Diffusion de la lumiere et des rayonnes X par un corps transparent homogene; influence del'agitation thermique," *Ann. Phys. (Paris)*, 17, 88–122, 1922.

[16] L. Mandelstam, "Light scattering by inhomogeneous media," *Zh. Russ. Fiz-Khim. Ova.*, 58, 381, 1926.

[17] E. Gross, "Change of Wave-length of Light due to Elastic Heat Waves at Scattering in Liquids." *Nature*, 126, 201–202, 1930.

[18] R. Y. Chiao, C. H. Townes, and B. P. Stoicheff, "Stimulated Brillouin Scattering and Coherent Generation of Intense Hypersonic Waves," *Phys. Rev. Lett.*, 12, 592–595, 1964.

[19] P. Kaiser and H. W. Astle, "Low-Loss Single-Material Fibers Made From Pure Fused Silica," *Bell System Technical Journal*, 53(6), 1021–1039, 1974.

[20] S. Miller, E. Marcatili, and T. Li, "Research toward optical-fiber transmission systems," *Proc. IEEE*, 61(12), 1703–1751, 1973.

[21] E. Ippen and R. Stolen, "Stimulated Brillouin scattering in optical fibers," *Appl. Phys. Lett.*, 21(11), 539–541, 1972.

[22] R. G. Smith, "Optical Power Handling Capacity of Low Loss Optical Fibers as Determined by Stimulated Raman and Brillouin Scattering," *Appl. Opt.*, 11(11), 2489–2494, 1972.

[23] D. Cotter, "Stimulated Brillouin scattering in monomode optical fiber," *J. Opt. Commun.*, 1(4), 10–19, 1983.

[24] A. Kobyakov, M. Sauer, and D. Chowdhury, "Stimulated Brillouin scattering in optical fibers," *Adv. Opt. Photon.*, 2(1), 1–59, 2010.

[25] Y. Aoki, K. Tajima, and I. Mito, "Input power limits of single-mode optical fibers due to stimulated Brillouin scattering in optical communication systems," *J. Lightw. Technol.*, 6(5), 710–719, 1988.

[26] A. Hirose, . Takushima, and T. Okoshi, "Suppression of Stimulated Brillouin Scattering and Brillouin Crosstalk by Frequency-Sweeping Spread-Spectrum Scheme," *J. Opt. Commun.*, 12(3), 82–85, 1991.

[27] F. H. Tithi and M. S. Islam, "Suppression of stimulated Brillouin scattering effect using nonlinear phase modulation," in *2010 International Conference on Electrical and Computer Engineering (ICECE)*, pages 135–138, 2010.

[28] H. Lee and G. P. Agrawal, "Suppression of stimulated Brillouin scattering in optical fibers using fiber Bragg gratings," *Opt. Express*, 11(25), 3467–3472, 2003.

[29] Y. Takushima and T. Okoshi, "Suppression of stimulated Brillouin scattering using isolators," in *Digest of Conference on Optical Fiber Communication*, page WM6, Optical Society of America, 1992.

[30] A. Kobyakov, S. Kumar, D. Q. Chowdhury, et al., "Design concept for optical fibers with enhanced SBS threshold," *Opt. Express*, 13(14), 5338–5346, 2005.

[31] T. Sugie, "Suppression of SBS by discontinuous Brillouin frequency shifted fibre in CPFSK coherent lightwave system with booster amplifier," *Electron. Lett.*, 27, 1231–1233(2), 1991.

[32] T. Kurashima, T. Horiguchi, and M. Tateda, "Distributed-temperature sensing using stimulated Brillouin scattering in optical silica fibers," *Opt. Lett.*, 15(18), 1038–1040, 1990.

[33] T. Horiguchi, K. Shimizu, T. Kurashima, et al., "Development of a distributed sensing technique using Brillouin scattering," *J. Lightw. Technol.*, 13(7), 1296–1302, 1995.

[34] C. A. Galindez-Jamioy and J. M. López-Higuera, "Brillouin distributed fiber sensors: an overview and applications," *Journal of Sensors*, 2012, 2012.

[35] S. P. Smith, F. Zarinetchi, and S. Ezekiel, "Narrow-linewidth stimulated Brillouin fiber laser and applications," *Opt. Lett.*, 16(6), 393–395, 1991.

[36] J. Geng, S. Staines, Z. Wang, et al., "Highly stable low-noise Brillouin fiber laser with ultranarrow spectral linewidth," *IEEE Photon. Technol. Lett.*, 18(17), 1813–1815, 2006.

[37] J. Xu, X. Ren, W. Gong, et al., "Measurement of the bulk viscosity of liquid by Brillouin scattering," *Appl. Opt.*, 42(33), 6704–6709, 2003.

[38] X. Ren, Z. Tian, Y. Zhang, et al., "Theoretical and experimental investigations on measuring underwater temperature by the coherent Brillouin scattering method," *Appl. Opt.*, 54(30), 9025–9029, 2015.

[39] C. W. Ballmann, J. V. Thompson, A. J. Traverso, et al., "Stimulated Brillouin Scattering Microscopic Imaging," *Sci. Rep.*, 5, 2015.

[40] S. Speziale, H. Marquardt, and T. S. Duffy, "Brillouin Scattering and its Application in Geosciences," *Rev. Mineral. Geochem.*, 78(1), 543–603, 2014.

[41] M. J. Damzen, V. Vlad, A. Mocofanescu, et al., *Stimulated Brillouin scattering: fundamentals and applications*, CRC press, 2003.

[42] T. Schneider, *Nonlinear Optics in Telecommunications*, Heidelberg: Springer-Verlag, Berlin, 2004.

[43] M. O. Van Deventer, "Polarization properties of Rayleigh backscattering in single-mode fibers," *J. Lightw. Technol.*, 11(12), 1895–1899, 1993.

[44] M. van Deventer and A. Boot, "Polarization properties of stimulated Brillouin scattering in single-mode fibers," *J. Lightw. Technol.*, 12(4), 585–590, 1994.

[45] R. H. Stolen, "Polarization effects in fiber Raman and Brillouin lasers," *J. Quantum Electron.*, 15, 1157–1160, 1979.

[46] A. Zadok, E. Zilka, A. Eyal, et al., "Vector analysis of stimulated Brillouin scattering

amplification in standard single-mode fibers," *Opt. Express*, 16(26), 21692–21707, 2008.

[47] A. Wise, M. Tur, and A. Zadok, "Sharp tunable optical filters based on the polarization attributes of stimulated Brillouin scattering," *Opt. Express*, 19(22), 21945–21955, 2011.

[48] A. Galtarossa, L. Palmieri, M. Santagiustina, et al., "Polarized Brillouin amplification in randomly birefringent and unidirectionally spun fibers," *IEEE Photon. Technol. Lett.*, 16(20), 1420–1422, 2008.

[49] L. Ursini, M. Santagiustina, and L. Palmieri, "Polarization-dependent Brillouin gain in randomly birefringent fibers," *IEEE Photon. Technol. Lett.*, 22(10), 712–714, 2010.

[50] O. Shlomovits and M. Tur, "Vector analysis of depleted stimulated Brillouin scattering amplification in standard single-mode fibers with nonzero birefringence," *Opt. Lett.*, 38(6), 836–838, 2013.

[51] W. B. Gardner, "Appendix on Nonlinearities for G.650," I,T U Document COM 15-273-E, 1996.

[52] T. C. E. Jones, *The validity of the single mode optical fibre transfer standard for the calibration of fibres for high power users*, Nat. Phys. Lab., 1998.

[53] K. Shiraki, M. Ohashi, and M. Tateda, "SBS threshold of a fiber with a Brillouin frequency shift distribution," *J. Lightw. Technol.*, 14(1), 50–57, 1996.

[54] R. Billington, "Measurement Methods for Stimulated Raman and Brillouin Scattering in Optical Fibres," NPL Report COEM 31, 1999.

[55] P. Bayvel and P. Radmore, "Solutions of the SBS equations in single mode optical fibres and implications for fibre transmission systems," *Electron. Lett.*, 26(7), 434–436, 1990.

[56] R. Esman and K. Williams, "Brillouin scattering: beyond threshold," in *Optical Fiber Communications, 1996. OFC '96*, pages 227–228, 1996.

[57] L. Thévenaz, "Limitations Caused by Nonlinear Effects," *COST 241 – Final Report*, pages 97–100, 1998.

[58] S. L. Floch and P. Cambon, "Theoretical evaluation of the Brillouin threshold and the steady-state Brillouin equations in standard single-mode optical fibers," *J. Opt. Soc. Am. A*, 20(6), 1132–1137, 2003.

[59] A. Kobyakov, M. Mehendale, M. Vasilyev, et al., "Stimulated Brillouin scattering in Raman-pumped fibers: a theoretical approach," *J. Lightw. Technol.*, 20(8), 1635–1643, 2002.

[60] J. Limpert, F. Roser, S. Klingebiel, et al., "The Rising Power of Fiber Lasers and Amplifiers," *IEEE J. Sel. Topics Quatum Electron.*, 13(3), 537–545, 2007.

[61] T. Y. Fan, "Laser beam combining for high-power, high-radiance sources," *IEEE J. Sel. Topics Quatum Electron.*, 11(3), 567–577, 2005.

[62] J. E. Rothenberg, P. A. Thielen, M. Wickham, et al., "Suppression of stimulated Brillouin scattering in single-frequency multi-kilowatt fiber amplifiers," in *Lasers and Applications in Science and Engineering*, pages 687300–687300, International Society for Optics and Photonics, 2008.

[63] N. Olsson and J. Van Der Ziel, "Fibre Brillouin amplifier with electronically controlled bandwidth," *Electron. Lett.*, 9(22), 488–490, 1986.

[64] C. Atkins, D. Cotter, D. Smith, et al., "Application of Brillouin amplification in coherent optical transmission," *Electron. Lett.*, 22, 556–558, 1986.

[65] Y. Stern, K. Zhong, T. Schneider, et al., "Tunable sharp and highly selective microwave-photonic band-pass filters based on stimulated Brillouin scattering," *Photonics Research*, 2(4), B18–B25, 2014.

[66] A. R. Chraplyvy and R. W. Tkach, "Narrowband tunable optical filter for channel selection in densely packed WDM systems," *Electron. Lett.*, 22(20), 1084–1085, 1986.

[67] K. Hill, B. Kawasaki, and D. Johnson, "CW Brillouin laser," *Appl. Phys. Lett.*, 28(10), 608–609, 1976.

[68] S. Smith, F. Zarinetchi, and S. Ezekiel, "Narrow-linewidth stimulated Brillouin fiber laser and applications," *Opt. Lett.*, 16(6), 393–395, 1991.

[69] J. Geng, S. Staines, Z. Wang, et al., "Highly stable low-noise Brillouin fiber laser with ultranarrow spectral linewidth," *IEEE Photonics Technology Letters*, 18(17), 1813–1815, 2006.

[70] F. Zarinetchi, S. Smith, and S. Ezekiel, "Stimulated Brillouin fiber-optic laser gyroscope," *Opt. Lett.*, 16(4), 229–231, 1991.

[71] M. Nikles, L. Thévenaz, and P. A. Robert, "Simple distributed fiber sensor based on Brillouin gain spectrum analysis," *Opt. Lett.*, 21(10), 758–760, 1996.

[72] X. Bao, D. J. Webb, and D. A. Jackson, "32-km distributed temperature sensor based on Brillouin loss in an optical fiber," *Opt. Lett.*, 18(18), 1561–1563, 1993.

[73] X. Bao, D. J. Webb, and D. A. Jackson, "Combined distributed temperature and strain sensor based on Brillouin loss in an optical fiber," *Opt. Lett.*, 19(2), 141–143, 1994.

[74] X. Bao and L. Chen, "Recent progress in Brillouin scattering based fiber sensors," *Sensors*, 11(4), 4152–4187, 2011.

[75] K. Y. Song, S. Chin, N. Primerov, et al., "Time-Domain Distributed Fiber Sensor With 1 cm Spatial Resolution Based on Brillouin Dynamic Grating," *J. Lightw. Technol.*, 28(14), 2062–2067, 2010.

[76] Y. Dong, L. Chen, and X. Bao, "Extending the sensing range of Brillouin optical time-domain analysis combining frequency-division multiplexing and in-line EDFAs," *J. Lightw. Technol.*, 30(8), 1161–1167, 2012.

[77] J. Capmany, B. Ortega, and D. Pastor, "A tutorial on microwave photonic filters," *J. Lightw. Technol.*, 24(1), 201–229, 2006.

[78] B. Vidal, M. Piqueras, and J. Marti, "Tunable and reconfigurable photonic microwave filter based on stimulated Brillouin scattering," *Opt. Lett.*, 32(1), 23–25, 2007.

[79] W. Zhang and R. A. Minasian, "Widely tunable single-passband microwave photonic filter based on stimulated Brillouin scattering," *IEEE Photon. Technol. Lett.*, 23(23), 1775–1777, 2011.

[80] W. Zhang and R. A. Minasian, "Ultrawide tunable microwave photonic notch filter based on stimulated Brillouin scattering," *IEEE Photon. Technol. Lett.*, 24(14), 1182–1184, 2012.

[81] X. Han, L. Wang, Y. Shao, et al., "Filtering properties of the tunable microwave photonic filter with stimulated Brillouin scattering," *Opt. Eng.*, 53(6), 066110–066110, 2014.

[82] W. T. Wang, J. G. Liu, H. K. Mei, et al., "Microwave photonic filter with complex coefficient based on optical carrier phase shift utilizing two stimulated Brillouin scattering pumps," *IEEE Photon. J.*, 7(1), 1–8, 2015.

[83] L. Yi, W. Wei, Y. Jaouen, et al., "Ideal rectangular microwave photonic filter with high selectivity based on stimulated Brillouin scattering," in *Optical Fiber Communication Conference*, pages Tu3F–5, Optical Society of America, 2015.

[84] X. Bao, J. Dhliwayo, N. Heron, et al., "Experimental and theoretical studies on a distributed temperature sensor based on Brillouin scattering," *J. Lightw. Technol.*, 13(7), 1340–1348, 1995.

[85] T. Parker, M. Farhadiroushan, V. Handerek, et al., "A fully distributed simultaneous strain and temperature sensor using spontaneous Brillouin backscatter," *IEEE Photon. Technol. Lett.*, 9(7), 979–981, 1997.

[86] M. Nikles, L. Thévenaz, and P. A. Robert, "Brillouin gain spectrum characterization in single-mode optical fibers," *J. Lightw. Technol.*, 15(10), 1842–1851, 1997.

[87] S. Le Floch and P. Cambon, "Study of Brillouin gain spectrum in standard single-mode optical fiber at low temperatures (1.4–370 K) and high hydrostatic pressures (1–250 bars)," *Opt. Commun.*, 219(1), 395–410, 2003.

[88] S. Le Floch, F. Riou, and P. Cambon, "Experimental and theoretical study of the Brillouin linewidth and frequency at low temperature in standard single-mode optical fibres," *J. Opt. A – Pure Appl. Op.*, 3(3), L12, 2001.

[89] A. S. Pine, "Brillouin scattering study of acoustic attenuation in fused quartz," *Phys. Rev.*, 185(3), 1187, 1969.

[90] J. Jäckle, "On the ultrasonic attenuation in glasses at low temperatures," *Zeitschrift*

für Physik, 257(3), 212–223, 1972.

[91] A. Fellay, L. Thévenaz, J. P. Garcia, et al., "Brillouin-based temperature sensing in optical fibres down to 1 k," in *Optical Fiber Sensors Conference Technical Digest, 2002. Ofs 2002, 15th*, pages 301–304, IEEE, 2002.

[92] A. Yeniay, J.-M. Delavaux, and J. Toulouse, "Spontaneous and stimulated Brillouin scattering gain spectra in optical fibers," *J. Lightw. Technol.*, 20(8), 1425 – 1432, 2002.

[93] R. W. Tkach, A. R. Chraplyvy, and R. Derosier, "Spontaneous Brillouin scattering for single-mode optical-fibre characterisation," *Electron. Lett.*, 22(19), 1011–1013, 1986.

[94] N. Shibata, R. G. Waarts, and R. P. Braun, "Brillouin-gain spectra for single-mode fibers having pure-silica, GeO_2-doped, and P_2O_5-doped cores," *Opt. Lett.*, 12(4), 269–271, 1987.

[95] P. D. Dragic, "Estimating the effect of Ge doping on the acoustic damping coefficient via a highly Ge-doped MCVD silica fiber," *J. Opt. Soc. Am. B*, 26(8), 1614–1620, 2009.

[96] P. Wait and T. Newson, "Measurement of Brillouin scattering coherence length as a function of pump power to determine Brillouin linewidth," *Optics communications*, 117(1), 142–146, 1995.

[97] R. W. Boyd, K. Rzaewski, and P. Narum, "Noise initiation of stimulated Brillouin scattering," *Phys. Rev. A*, 42, 5514–5521, 1990.

[98] P. Wait, *The application of Brillouin scattering to distributed fibre optic sensing*, Ph.D. thesis, University of Southampton, 1997.

[99] S. Chi, C. Lee, P. Chiang, et al., "Measurement of stimulated-Brillouin-scattering thresholds for various types of fibers using Brillouin optical-time-domain reflectometer," in *Lasers and Electro-Optics, 2000. (CLEO 2000). Conference on*, pages 305–306, 2000.

[100] L. Xing, L. Zhan, L. Yi, et al., "Storage capacity of slow-light tunable optical buffers based on fiber Brillouin amplifiers for real signal bit streams," *Opt. Express*, 15(16), 10189–10195, 2007.

[101] S. Preußler and T. Schneider, "Bandwidth reduction in a multistage Brillouin system," *Opt. Lett.*, 37(19), 4122–4124, 2012.

[102] A. L. Gaeta and R. W. Boyd, "Stimulated Brillouin scattering in the presence of external feedback," *International Journal of Nonlinear Optical Physics*, 1(03), 581–594, 1992.

[103] S. Preussler and T. Schneider, "Stimulated Brillouin scattering gain bandwidth reduction and applications in microwave photonics and optical signal processing," *Opt. Eng.*, 55(3), 031110–031110, 2016.

[104] A. Wiatrek, S. Preußler, K. Jamshidi, et al., "Frequency domain aperture for the gain bandwidth reduction of stimulated Brillouin scattering," *Opt. Lett.*, 37(5), 930–932,

2012.

[105] A. Wiatrek, *Untersuchung der Eigenschaften der gesättigten stimulierten Brillouin-Streuung und ihrer Anwendungsmöglichkeiten*, Ph.D. thesis, Brandenburgische Technische Universität Cottbus, 2014.

[106] S. Preußler, A. Wiatrek, K. Jamshidi, et al., "Brillouin scattering gain bandwidth reduction down to 3.4 MHz," *Opt. Express*, 19(9), 8565–8570, 2011.

[107] T. Schneider, R. Henker, K.-U. Lauterbach, et al., "Comparison of delay enhancement mechanisms for SBS-based slow light systems," *Opt. Express*, 15(15), 9606–9613, 2007.

[108] T. Schneider, R. Henker, K.-U. Lauterbach, et al., "Distortion reduction in Slow Light systems based on stimulated Brillouin scattering," *Opt. Express*, 16(11), 8280–8285, 2008.

[109] Agilent, "Application Note 1550-4 – Optical Spectrum Analysis," 2000.

[110] E. Voges and K. Petermann, *Optische Kommunikationstechnik: Handbuch für Wissenschaft und Industrie*, Springer, Berlin, 2002.

[111] D. M. Baney, B. Szafraniec, and A. Motamedi, "Coherent optical spectrum analyzer," *IEEE Photon. Technol. Lett.*, 14(3), 355–357, 2002.

[112] F. R. Giorgetta, I. Coddington, E. Baumann, et al., "Fast high-resolution spectroscopy of dynamic continuous-wave laser sources," *Nature Photon.*, 4(12), 853–857, 2010.

[113] J. Domingo, J. Pelayo, F. Villuendas, et al., "Very high resolution optical spectrometry by stimulated Brillouin scattering," *IEEE Photon. Technol. Lett.*, 17(4), 855–857, 2005.

[114] T. Schneider, "Wavelength and line width measurement of optical sources with femtometre resolution," *Electron. Lett.*, 41(22), 1234–1235, 2005.

[115] S. Treff, S. Preussler, and T. Schneider, "Measuring the spectra of advanced optical signals with an extension of an electrical network analyzer," in *Optical Fiber Communication Conference and Exposition and the National Fiber Optic Engineers Conference (OFC/NFOEC), 2013*, pages 1–3, 2013.

[116] Y. Ma, Q. Yang, Y. Tang, et al., "1-Tb/s single-channel coherent optical OFDM transmission with orthogonal-band multiplexing and subwavelength bandwidth access," *J. Lightw. Technol.*, 28(4), 308–315, 2010.

[117] D. Hillerkuss, R. Schmogrow, T. Schellinger, et al., "26 Tbit s-1 line-rate super-channel transmission utilizing all-optical fast Fourier transform processing," *Nature Photon.*, 5(6), 364–371, 2011.

[118] T. Kuri, H. Toda, J. J. V. Olmos, et al., "Reconfigurable dense wavelength-division-multiplexing millimeter-waveband radio-over-fiber access system technologies," *J. Lightw. Technol.*, 28(16), 2247–2257, 2010.

[119] C. S. Park, Y.-K. Yeo, and L. C. Ong, "Demonstration of the GbE service in the

converged radio-over-fiber/optical networks," *J. Lightw. Technol.*, 28(16), 2307–2314, 2010.

[120] T. Schneider, A. Wiatrek, S. Preußler, et al., "Link budget analysis for terahertz fixed wireless links," *IEEE Trans. THz Sci. Technol.*, 2(2), 250–256, 2012.

[121] I. M. White and X. Fan, "On the performance quantification of resonant refractive index sensors," *Opt. Express*, 16(2), 1020–1028, 2008.

[122] A. M. Armani, R. P. Kulkarni, S. E. Fraser, et al., "Label-free, single-molecule detection with optical microcavities," *Science*, 317(5839), 783–787, 2007.

[123] F. Mihélic, D. Bacquet, J. Zemmouri, et al., "Ultrahigh resolution spectral analysis based on a Brillouin fiber laser," *Opt. Lett.*, 35(3), 432–434, 2010.

[124] K. Y. Song, K. Lee, and S. B. Lee, "Tunable optical delays based on Brillouin dynamic grating in optical fibers," *Opt. Express*, 17(12), 10344–10349, 2009.

[125] S. Preussler, A. Zadok, A. Wiatrek, et al., "Enhancement of spectral resolution and optical rejection ratio of Brillouin optical spectral analysis using polarization pulling," *Opt. Express*, 20(13), 14734–14745, 2012.

[126] S. Preussler, A. Wiatrek, K. Jamshidi, et al., "Ultrahigh-resolution spectroscopy based on the bandwidth reduction of stimulated Brillouin scattering," *IEEE Photon. Technol. Lett.*, 23(16), 1118–1120, 2011.

[127] S. Preussler, A. Wiatrek, K. Jamshidi, et al., "Increasing the resolution of optical spectrometers for the measurement of advanced optical communication signals," in *European Conference and Exhibition on Optical Communication*, pages We–1, Optical Society of America, 2012.

[128] J. Gordon and H. Kogelnik, "PMD fundamentals: Polarization mode dispersion in optical fibers," *P. Natl. Acad. Sci. USA*, 97(9), 4541–4550, 2000.

[129] A. Voskoboinik, J. Wang, B. Shamee, et al., "SBS-based fiber optical sensing using frequency-domain simultaneous tone interrogation," *J. Lightw. Technol.*, 29(11), 1729–1735, 2011.

[130] S. Preussler and T. Schneider, "Attometer resolution spectral analysis based on polarization pulling assisted Brillouin scattering merged with heterodyne detection," *Opt. Express*, 23(20), 26879–26887, 2015.

[131] S. Preussler, H. Al-Taiy, and T. Schneider, "Optical spectrum analysis with kHz resolution based on polarization pulling and local oscillator assisted Brillouin scattering," in *Optical Communication (ECOC), 2015 European Conference on*, pages 1–3, IEEE, 2015.

[132] R. S. Tucker, P.-C. Ku, and C. J. Chang-Hasnain, "Slow-Light Optical Buffers: Capabilities and Fundamental Limitations," *J. Lightw. Technol.*, 23(12), 4046, 2005.

[133] L. V. Hau, S. E. Harris, Z. Dutton, et al., "Light speed reduction to 17 metres per second in an ultracold atomic gas," *Nature*, 397(6720), 594–598, 1999.

[134] S. Fan and M. Yanik, "All-optical coherent stopping and storage of light," in *Lasers and Electro-Optics, 2004. (CLEO). Conference on*, vol. 1, page 2 pp. vol.1, 2004.

[135] H. Lin, T. Wang, and T. W. Mossberg, "Demonstration of 8-Gbit/in.^{2} areal storage density based on swept-carrier frequency-selective optical memory," *Opt. Lett.*, 20(15), 1658–1660, 1995.

[136] C. Liu, Z. Dutton, C. H. Behroozi, et al., "Observation of coherent optical information storage in an atomic medium using halted light pulses," *Nature*, 409, 490–493, 2001.

[137] J. J. Longdell, E. Fraval, M. J. Sellars, et al., "Stopped Light with Storage Times Greater than One Second Using Electromagnetically Induced Transparency in a Solid," *Phys. Rev. Lett.*, 95, 063601, 2005.

[138] L. Thévenaz, "Slow and fast light in optical fibres," *Nature Photon.*, 2, 474–481, 2008.

[139] Y. Okawachi, M. S. Bigelow, J. E. Sharping, et al., "Tunable All-Optical Delays via Brillouin Slow Light in an Optical Fiber," *Phys. Rev. Lett.*, 94, 153902, 2005.

[140] Z. Zhu, A. Dawes, D. Gauthier, et al., "Broadband SBS Slow Light in an Optical Fiber," *J. Lightw. Technol.*, 25(1), 201 –206, 2007.

[141] Z. Lu, Y. Dong, and Q. Li, "Slow light in multi-line Brillouin gain spectrum," *Opt. Express*, 15(4), 1871–1877, 2007.

[142] E. Cabrera-Granado, O. G. Calderon, S. Melle, et al., "Observation of large 10-Gb/s SBS slow light delay with low distortion using an optimized gain profile," *Opt. Express*, 16(20), 16032–16042, 2008.

[143] R. W. Boyd, D. J. Gauthier, A. L. Gaeta, et al., "Maximum time delay achievable on propagation through a slow-light medium," *Phys. Rev. A*, 71, 023801, 2005.

[144] T. Schneider, "Time delay limits of stimulated-Brillouin-scattering-based slow light systems," *Opt. Lett.*, 33(13), 1398–1400, 2008.

[145] A. Wiatrek, K. Jamshidi, R. Henker, et al., "Nonlinear Brillouin based slow-light system for almost distortion-free pulse delay," *J. Opt. Soc. Am. B*, 27(3), 544–549, 2010.

[146] Z. Hu, J. Sun, L. Liu, et al., "All-optical tunable delay line based on wavelength conversion and fiber dispersion," *Proc. SPIE 6838*, pages 68380N–68380N–8, 2007.

[147] O. Yilmaz, S. Nuccio, X. Wu, et al., "40-Gb/s Optical Packet Buffer Using Conversion/Dispersion-Based Delays," *J. Lightw. Technol.*, 28(4), 616 –623, 2010.

[148] Y. Okawachi, J. E. Sharping, C. Xu, et al., "Large tunable optical delays via self-phase modulation and dispersion," *Opt. Express*, 14(25), 12022–12027, 2006.

[149] J. Sharping, Y. Okawachi, J. van Howe, et al., "All-optical, wavelength and bandwidth preserving, pulse delay based on parametric wavelength conversion and dispersion,"

Opt. Express, 13(20), 7872–7877, 2005.

[150] T. Kurosu and S. Namiki, "Continuously tunable 22 ns delay for wideband optical signals using a parametric delay-dispersion tuner," *Opt. Lett.*, 34(9), 1441–1443, 2009.

[151] Y. Okawachi, M. A. Foster, X. Chen, et al., "Large tunable delays using parametric mixing and phase conjugation in Si nanowaveguides," *Opt. Express*, 16(14), 10349–10357, 2008.

[152] T. F. Krauss, "Why do we need slow light?" *Nature Photon.*, 2, 448–450, 2008.

[153] E. Burmeister, D. Blumenthal, and J. Bowers, "A comparison of optical buffering technologies," *Opt. Switch. Netw.*, 5(1), 10 – 18, 2008.

[154] J. LeGrange, J. Simsarian, P. Bernasconi, et al., "Demonstration of an Integrated Buffer for an All-Optical Packet Router," *IEEE Photon. Technol. Lett.*, 21(12), 781 –783, 2009.

[155] R. Langenhorst, M. Eiselt, W. Pieper, et al., "Fiber loop optical buffer," *J. Lightw. Technol.*, 14(3), 324–335, 1996.

[156] K. Y. Song, W. Zou, Z. He, et al., "All-optical dynamic grating generation based on Brillouin scattering in polarization-maintaining fiber," *Opt. Lett.*, 33(9), 926–928, 2008.

[157] Z. Zhu, D. J. Gauthier, and R. W. Boyd, "Stored Light in an Optical Fiber via Stimulated Brillouin Scattering," *Science*, 318(5857), 1748–1750, 2007.

[158] K. Jamshidi, S. Preussler, A. Wiatrek, et al., "A Review to the All-Optical Quasi-Light Storage," *IEEE J. Sel. Topics Quatum Electron.*, 18(2), 884–890, 2012.

[159] S. Preussler, K. Jamshidi, A. Wiatrek, et al., "Almost distortion free storage of 1Gbps/8bit optical packets for up to 100 bit lengths," in *Optical Communication (ECOC), 2010 36th European Conference and Exhibition on*, pages 1–3, IEEE, 2010.

[160] S. Preußler, K. Jamshidi, A. Wiatrek, et al., "Quasi-Light-Storage based on time-frequency coherence," *Opt. Express*, 17(18), 15790–15798, 2009.

[161] T. Sakamoto, T. Kawanishi, and M. Izutsu, "19x10-GHz Electro-Optic Ultra-Flat Frequency Comb Generation Only Using Single Conventional Mach-Zehnder Modulator," in *Conference on Lasers and Electro-Optics/Quantum Electronics and Laser Science Conference and Photonic Applications Systems Technologies*, page CMAA5, Optical Society of America, 2006.

[162] G. Sefler and K. Kitayama, "Frequency comb generation by four-wave mixing and the role of fiber dispersion," *J. Lightw. Technol.*, 16(9), 1596–1605, 1998.

[163] T. Schneider, K. Jamshidi, and S. Preussler, "Quasi-Light Storage: A Method for the Tunable Storage of Optical Packets With a Potential Delay-Bandwidth Product of Several Thousand Bits," *J. Lightw. Technol.*, 28(17), 2586–2592, 2010.

[164] C. Rulliere, *Femtosecond laser pulses*, Springer, 2005.

[165] T. Schneider, M. Junker, and K.-U. Lauterbach, "Theoretical and experimental investigation of Brillouin scattering for the generation of millimeter waves," *J. Opt. Soc. Am. B*, 23(6), 1012–1019, 2006.

[166] T. Schneider, M. Junker, and K.-U. Lauterbach, "Potential ultra wide slow-light bandwidth enhancement," *Opt. Express*, 14(23), 11082–11087, 2006.

[167] R. Henker, A. Wiatrek, K.-U. Lauterbach, et al., "Group velocity dispersion reduction in fibre-based slow-light systems via stimulated Brillouin scattering," *Electron. Lett.*, 44(20), 1185–1186, 2008.

[168] R. S. Tucker, P.-C. Ku, and C. J. Chang-Hasnain, "Delay-bandwidth product and storage density in slow-light optical buffers," *Electron. Lett.*, 41(4), 208–209, 2005.

[169] M. González Herráez, K. Y. Song, and L. Thévenaz, "Arbitrary-bandwidth Brillouin slow light in optical fibers," *Opt. Express*, 14(4), 1395–1400, 2006.

[170] T. Schneider and S. Preußler, "Quasi-light Storage for Optical Data Packets," *J. Vis. Exp.*, 84, e50468–e50468, 2014.

[171] A. H. Gnauck and P. J. Winzer, "Optical phase-shift-keyed transmission," *J. Lightw. Technol.*, 23(1), 115, 2005.

[172] S. Preußler and T. Schneider, "All optical tunable storage of phase-shift-keyed data packets," *Opt. Express*, 20(16), 18224–18229, 2012.

[173] B. Zhang, L. Yan, I. Fazal, et al., "Slow light on Gbit/s differential-phase-shift-keying signals," *Opt. Express*, 15(4), 1878–1883, 2007.

[174] S. Preussler and T. Schneider, "Proposal for the tunable all optical storage of QAM data packets," in *CLEO: Science and Innovations*, pages JTu4A–96, Optical Society of America, 2013.

[175] S. Preussler, A. Wiatrek, K. Jamshidi, et al., "Methods for the enhancement of the storage time in Quasi-Light-Storage," in *2011 International Students and Young Scientists Workshop "Photonics and Microsystems"*, 2011.

[176] S. Preussler, A. Wiatrek, K. Jamshidi, et al., "Quasi-light-storage enhancement by reducing the Brillouin gain bandwidth," *Appl. Opt.*, 50(22), 4252–4256, 2011.

[177] S. Preussler, K. Jamshidi, A. Wiatrek, et al., "Light Storage Enhancement by Reducing the Brillouin Bandwidth," in *Slow and Fast Light*, page SLWA4, Optical Society of America, 2011.

[178] S. Preussler, K. Jamshidi, A. Wiatrek, et al., "Very simple tunable optical data storage of 8Bit 1Gbps data packets up to 500ns," in *CLEO: Science and Innovations*, page CFP5, Optical Society of America, 2011.

[179] S. Preußler, K. Jamshidi, A. Wiatrek, et al., "Einfache, variable, optische Datenspeicherung bis zu 800 ns," *ITG-Fachbericht-Photonische Netze*, 2011.

[180] R. Pant, C. G. Poulton, D.-Y. Choi, et al., "On-chip stimulated Brillouin scattering," *Opt. Express*, 19(9), 8285–8290, 2011.

[181] J. Ye, H. Schnatz, and L. W. Hollberg, "Optical frequency combs: from frequency metrology to optical phase control," *IEEE J. Sel. Topics Quatum Electron.*, 9(4), 1041–1058, 2003.

[182] T. Udem, R. Holzwarth, and T. W. Hänsch, "Optical frequency metrology," *Nature*, 416(6877), 233–237, 2002.

[183] P. Maddaloni, P. Cancio, and P. De Natale, "Optical comb generators for laser frequency measurement," *Meas. Sci. Technol.*, 20(5), 052001, 2009.

[184] M. J. Thorpe, K. D. Moll, R. J. Jones, et al., "Broadband cavity ringdown spectroscopy for sensitive and rapid molecular detection," *Science*, 311(5767), 1595–1599, 2006.

[185] A. Marian, M. C. Stowe, J. R. Lawall, et al., "United time-frequency spectroscopy for dynamics and global structure," *Science*, 306(5704), 2063–2068, 2004.

[186] W. Oskay, S. Diddams, E. Donley, et al., "Single-atom optical clock with high accuracy," *Phys. Rev. Lett.*, 97(2), 020801, 2006.

[187] A. D. Ludlow, T. Zelevinsky, G. Campbell, et al., "Sr lattice clock at $1\times 10{-}16$ fractional uncertainty by remote optical evaluation with a Ca clock," *Science*, 319(5871), 1805–1808, 2008.

[188] Y. Okawachi, K. Saha, J. S. Levy, et al., "Octave-spanning frequency comb generation in a silicon nitride chip," *Opt. Lett.*, 36(17), 3398–3400, 2011.

[189] P. Del'Haye, A. Schliesser, O. Arcizet, et al., "Optical frequency comb generation from a monolithic microresonator," *Nature*, 450(7173), 1214–1217, 2007.

[190] T. Sakamoto, T. Kawanishi, and M. Izutsu, "Optimization of electro-optic comb generation using conventional Mach-Zehnder modulator," in *Microwave Photonics, 2007 IEEE International Topical Meeting on*, pages 50–53, IEEE, 2007.

[191] R. Wu, V. Supradeepa, C. M. Long, et al., "Generation of very flat optical frequency combs from continuous-wave lasers using cascaded intensity and phase modulators driven by tailored radio frequency waveforms," *Opt. Lett.*, 35(19), 3234–3236, 2010.

[192] T. Kobayashi, T. Sueta, Y. Cho, et al., "High-repetition-rate optical pulse generator using a Fabry-Perot electro-optic modulator," *Appl. Phys. Lett.*, 21(8), 341–343, 1972.

[193] T. Schneider, "Fiber-laser frequency combs for the generation of tunable single-frequency laser lines, mm- and THz-waves and sinc-shaped Nyquist pulses," in *Proc. SPIE*, vol. 9378, pages 937822–937822–8, 2015.

[194] H. Al-Taiy, N. Wenzel, S. Preußler, et al., "Ultra-narrow linewidth, stable and tunable laser source for optical communication systems and spectroscopy," *Opt. Lett.*, 39(20), 5826–5829, 2014.

[195] S. Preußler, N. Wenzel, R.-P. Braun, et al., "Generation of ultra-narrow, stable and tunable millimeter-and terahertz-waves with very low phase noise," *Opt. Express*, 21(20), 23950–23962, 2013.

[196] S. Preussler, N. Wenzel, A. Zadok, et al., "Tunable generation of ultra-narrow linewidth millimeter and THz-waves and their modulation at 40 Gbd," in *Microwave Photonics (MWP), 2013 International Topical Meeting on*, pages 119–122, IEEE, 2013.

[197] S. Preussler, N. Wenzel, and T. Schneider, "Flat, rectangular frequency comb generation with tunable bandwidth and frequency spacing," *Opt. Lett.*, 39(6), 1637–1640, 2014.

[198] S. Preussler, N. Wenzel, and T. Schneider, "Flexible Nyquist pulse sequence generation with variable bandwidth and repetition rate," *IEEE Photon. J.*, 6(4), 1–8, 2014.

[199] M. Tonouchi, "Cutting-edge terahertz technology," *Nature Photon.*, 1(2), 97–105, 2007.

[200] C. Jansen, S. Wietzke, O. Peters, et al., "Terahertz imaging: applications and perspectives," *Appl. Opt.*, 49(19), E48–E57, 2010.

[201] P. H. Siegel, "Terahertz technology in biology and medicine," in *Microwave Symposium Digest, 2004 IEEE MTT-S International*, vol. 3, pages 1575–1578, IEEE, 2004.

[202] S. L. Dexheimer, *Terahertz spectroscopy: principles and applications*, CRC press, 2007.

[203] P. J. Winzer and R.-J. Essiambre, "Advanced modulation formats for high-capacity optical transport networks," *J. Lightw. Technol.*, 24(12), 4711–4728, 2006.

[204] M. Feiginov, C. Sydlo, O. Cojocari, et al., "Resonant-tunnelling-diode oscillators operating at frequencies above 1.1 THz," *Appl. Phys. Lett.*, 99(23), 233506, 2011.

[205] P. H. Siegel, "Terahertz technology," *IEEE Trans. Microw. Theory Techn.*, 50(3), 910–928, 2002.

[206] J. C. Pearson, B. J. Drouin, A. Maestrini, et al., "Demonstration of a room temperature 2.48–2.75 THz coherent spectroscopy source," *Rev. Sci. Instrum.*, 82(9), 093105, 2011.

[207] I. C. Mayorga, A. Schmitz, T. Klein, et al., "First in-field application of a full photonic local oscillator to terahertz astronomy," *IEEE Trans. THz Sci. Technol.*, 2(4), 393–399, 2012.

[208] S. Ginestar, F. van Dijk, A. Accard, et al., "Tunable dual-mode DFB laser for millimetre-wave signal generation," *Eur. Phys. J. - Appl. Phys.*, 53(03), 33609, 2011.

[209] E. D. Black, "An introduction to Pound–Drever–Hall laser frequency stabilization," *Am. J. Phys.*, 69(1), 79–87, 2001.

[210] Q. Quraishi, M. Griebel, T. Kleine-Ostmann, et al., "Generation of phase-locked and tunable continuous-wave radiation in the terahertz regime," *Opt. Lett.*, 30(23), 3231–3233, 2005.

[211] D. Stanze, A. Deninger, A. Roggenbuck, et al., "Compact cw terahertz spectrometer pumped at 1.5 μm wavelength," *J. Infrared Millim. Terahertz Waves*, 32(2), 225–232,

2011.

[212] S. Preussler, T. Schneider, and H. Al-Taiy, "Generation and Stabilization of THz-waves with Extraordinary Low Line Width and Phase Noise," in *CLEO: Science and Innovations*, pages STu4H–6, Optical Society of America, 2015.

[213] G. Bosco, V. Curri, A. Carena, et al., "On the performance of Nyquist-WDM terabit superchannels based on PM-BPSK, PM-QPSK, PM-8QAM or PM-16QAM subcarriers," *J. Lightw. Technol.*, 29(1), 53–61, 2011.

[214] I. B. Djordjevic and B. Vasic, "Orthogonal frequency division multiplexing for high-speed optical transmission," *Opt. Express*, 14(9), 3767–3775, 2006.

[215] D. Hillerkuss, R. Schmogrow, M. Meyer, et al., "Single-laser 32.5 Tbit/s Nyquist WDM transmission," *IEEE J. Opt. Commun. Netw.*, 4(10), 715–723, 2012.

[216] G. Bosco, A. Carena, V. Curri, et al., "Performance limits of Nyquist-WDM and CO-OFDM in high-speed PM-QPSK systems," *IEEE Photon. Technol. Lett.*, 22(15), 1129–1131, 2010.

[217] R. Schmogrow, D. Hillerkuss, S. Wolf, et al., "512QAM Nyquist sinc-pulse transmission at 54 Gbit/s in an optical bandwidth of 3 GHz," *Opt. Express*, 20(6), 6439–6447, 2012.

[218] G. C. Valley, "Photonic analog-to-digital converters," *Opt. Express*, 15(5), 1955–1982, 2007.

[219] V. Supradeepa, C. M. Long, R. Wu, et al., "Comb-based radiofrequency photonic filters with rapid tunability and high selectivity," *Nature Photon.*, 6(3), 186–194, 2012.

[220] M. Song, C. M. Long, R. Wu, et al., "Reconfigurable and tunable flat-top microwave photonic filters utilizing optical frequency combs," *IEEE Photon. Technol. Lett.*, 23(21), 1618–1620, 2011.

[221] E. Hamidi, D. E. Leaird, and A. Weiner, "Tunable programmable microwave photonic filters based on an optical frequency comb," *IEEE Trans. Microw. Theory Techn.*, 58(11), 3269–3278, 2010.

[222] R. Schmogrow, M. Winter, M. Meyer, et al., "Real-time Nyquist pulse generation beyond 100 Gbit/s and its relation to OFDM," *Opt. Express*, 20(1), 317–337, 2012.

[223] M. Nakazawa, T. Hirooka, P. Ruan, et al., "Ultrahigh-speed "orthogonal" TDM transmission with an optical Nyquist pulse train," *Opt. Express*, 20(2), 1129–1140, 2012.

[224] A. Vedadi, M. A. Shoaie, and C.-S. Brès, "Near-Nyquist optical pulse generation with fiber optical parametric amplification," *Opt. Express*, 20(26), B558–B565, 2012.

[225] M. A. Soto, M. Alem, M. A. Shoaie, et al., "Optical sinc-shaped Nyquist pulses of exceptional quality," *Nat. Commun.*, 4, 2013.

[226] S. Preussler, N. Wenzel, and T. Schneider, "Generation of Flat, Rectangular Frequency

Combs with Tunable Bandwidth and Frequency Spacing," in *CLEO: Science and Innovations*, pages STu3I-2, Optical Society of America, 2014.

[227] E. P. da Silva, R. Borkowski, S. Preussler, et al., "Combined Optical and Electrical Spectrum Shaping for High-Baud-Rate Nyquist-WDM Transceivers," *IEEE Photon. J.*, 8(1), 1–11, 2016.

[228] H. Al-Taiy, S. Preussler, S. Brückner, et al., "Generation of highly stable millimeter waves with low phase noise and narrow linewidth," in *Photonics Conference (IPC), 2015*, pages 98–101, IEEE, 2015.

[229] H. Al-Taiy, S. Preussler, S. Brückner, et al., "Generation of highly stable millimeter waves with low phase noise and narrow linewidth," in *Photonics Conference (IPC), 2015*, pages 98–101, IEEE, 2015.

[230] H. Shin, W. Qiu, R. Jarecki, et al., "Tailorable stimulated Brillouin scattering in nanoscale silicon waveguides," *Nat. Commun.*, 4, 2013.

[231] R. Pant, C. G. Poulton, D.-Y. Choi, et al., "On-chip stimulated Brillouin scattering," *Opt. Express*, 19(9), 8285–8290, 2011.

[232] R. Pant, A. Byrnes, E. Li, et al., "Photonic chip based tunable and dynamically reconfigurable microwave photonic filter using stimulated Brillouin scattering," in *Nonlinear Photonics*, pages JW4D-5, Optical Society of America, 2012.

[233] R. Pant, A. Byrnes, C. G. Poulton, et al., "Photonic-chip-based tunable slow and fast light via stimulated Brillouin scattering," *Opt. Lett.*, 37(5), 969–971, 2012.

[234] R. Pant, D. Marpaung, I. V. Kabakova, et al., "On-chip stimulated Brillouin Scattering for microwave signal processing and generation," *Laser Photon. Rev.*, 8(5), 653–666, 2014.

[235] I. V. Kabakova, R. Pant, D.-Y. Choi, et al., "Narrow linewidth Brillouin laser based on chalcogenide photonic chip," *Opt. Lett.*, 38(17), 3208–3211, 2013.

[236] H. Lüke, *Signalübertragung*, Springer, 1999.

[237] N. Fliege, *Systemtheorie*, B.G. Teubner Verlag, 1991.

[238] W. S. Chang, *RF photonic technology in optical fiber links*, Cambridge University Press, 2002.

Due to priority reasons, parts of the results of this thesis have been already published in peer-reviewed scientific journals and on internationally renowned conferences and workshops.

Journals

[1] A. Wiatrek, R. Henker, S. Preußler, M. J. Ammann, A. T. Schwarzbacher, T. Schneider, "Zero-broadening measurement in Brillouin based slow-light delays," Opt. Express 17(2), 797–802 (2009).

[2] A. Wiatrek, R. Henker, S. Preußler, T. Schneider, "Pulse broadening cancellation in cascaded slow-light delays," Opt. Express 17(9), 7586–7591 (2009).

[3] S. Preußler, K. Jamshidi, A. Wiatrek, R. Henker, T. Schneider, "Quasi-Light-Storage based on time-frequency coherence," Opt. Express 17(18), 15790–15798 (2009).

[4] R. Henker, A. Wiatrek, S. Preußler, M. J. Ammann, A. T. Schwarzbacher, T. Schneider, "Gain enhancement in multiple-pump-line Brillouin-based slow light systems by using fiber segments and filter stages," Appl. Optics 48(29), 5583–5588 (2009).

[5] A. Wiatrek, K. Jamshidi, R. Henker, S. Preußler, T. Schneider, "Nonlinear Brillouin based slow-light system for almost distortion-free pulse delay," J. Opt. Soc. Am. B 27(3), 544–549 (2010).

[6] Thomas Schneider, K. Jamshidid, S. Preußler, "Quasi-Light Storage: A Method for the Tunable Storage of Optical Packets With a Potential Delay-Bandwidth Product of Several Thousand Bits," J. Lightwave Technol. 28(17), 2586–2592, (2010).

[7] K. Jamshidi, S. Preussler, A. Wiatrek, T. Schneider, "A review to the all optical Quasi Light Storage," IEEE J. Sel. Topics in Quantum Electron. 18(2), 884–890 (2011).

[8] S. Preussler, A. Wiatrek, K. Jamshidi, T. Schneider, "Brillouin Scattering Gain Bandwidth Reduction Down to 3.4 MHz," Opt. Express 19(9), 8565–8570 (2011).

[9] S. Preussler, A. Wiatrek, K. Jamshidi, T. Schneider, "Ultra-High Resolution Spectroscopy Based on the Bandwidth Reduction of Stimulated Brillouin Scattering," IEEE Photon. Technol. Lett. 23(16), 1118–1120 (2011).

[10] S. Preussler, A. Wiatrek, K. Jamshidi, T. Schneider, "Quasi-Light-Storage Enhancement by Reducing the Brillouin Gain Bandwidth," Appl. Optics 50(22), 4252–4256 (2011).

[11] S. Preußler, T. Schneider, "Auswirkungen der Umgebung auf die Ausbreitung elektromagnetischer Wellen," Wissen Heute 4/2011

[12] A. Wiatrek, S. Preußler, T. Schneider, "Voruntersuchung des Slow- und Fast-Light Effekts auf der Basis von stimulierter Brillouin Streuung für die Anwendbarkeit in optischen Kommunikations- und Informationssystemen," Schlussbericht des Wissenschaftlichen Vorprojektes (WiVorPro) in den Optischen Technologien, BMBF-Förderkennzeichen 13N9355, Technische Informationsbibliothek TIB Hannover, 2011.

[13] T. Schneider, A. Wiatrek, S. Preußler, M. Grigat, R.P. Braun, "Link Budget analysis for Terahertz Fixed Wireless Links," IEEE Trans. THz Sci. Technol. 2(2), 250–256 (2012).

[14] A. Wiatrek, S. Preussler, K. Jamshidi and T. Schneider, "Frequency Domain Aperture for the Gain Bandwidth Reduction of Stimulated Brillouin Scattering," Opt. Lett. 37(5), 930–932 (2012).

[15] S. Preussler, A. Zadok, A. Wiatrek, M. Tur, T. Schneider, "Enhancement of spectral resolution and optical rejection ratio of Brillouin optical spectral analysis using polarization pulling," Opt. Express 20(13), 14734–14745 (2012).

[16] S. Preußler, T. Schneider, "All Optical Tunable Storage of Phase-Shift-Keyed Data Packets," Opt. Express 20(16), 18224–18229 (2012).

[17] A. Mokhtari, S. Preußler, K. Jamshidi, M. Akbari, T. Schneider, "Fully-tunable microwave photonic filter with complex coefficients using tunable delay lines based on frequency-time conversions," Opt. Express 20(20), 22728–22734 (2012).

[18] S. Preußler, T. Schneider, "Bandwidth reduction in a multi-stage Brillouin system," Opt. Lett. 37(19), 4122–4124 (2012).

[19] S. Preußler, N. Wenzel, R.-P. Braun, N. Owschimikow, C. Vogel, A. Deninger, A. Zadok, U. Woggon, T. Schneider, "Generation of ultra-narrow, stable and tunable millimeter- and terahertz- waves with very low phase noise," Opt. Express 21(20), 23950–23962 (2013).

[20] A. Mokhtari, K. Jamshidi, S. Preußler, A. Zadok, T. Schneider, "Tunable Microwave-Photonic Filter Using Frequency-to-Time Mapping-Based Delay lines," Opt. Express 21(18), 21702–21707 (2013).

[21] T. Schneider, S. Preußler, "Quasi-Light-Storage for Optical Data Packets," J. Vis. Exp. (84), e50468, doi:10.3791/50468 (2014).

[22] S. Preussler, N. Wenzel, T. Schneider, "Flat Rectangular Frequency Comb Generation with Tunable Bandwidth and Frequency Distance," Opt. Lett. 39(6), 1637–1640 (2014).

[23] S. Preussler, N. Wenzel and T. Schneider, "Flexible Nyquist Pulse Sequence Generation with Variable Bandwidth and Repetition Rate," IEEE Photon. J. 6(4), 7901608 (2014).

[24] H. Al-Taiy, N. Wenzel, S. Preussler, J. Klinger, T. Schneider, "Ultra Narrow linewidth, Stable and Tunable Laser Source for Optical Communications Systems and Spectroscopy," Opt. Lett. 39(20), 5826–5829 (2014).

[25] H. Al-Taiy, S. Preussler, S. Brueckner, J. Schoebel, T. Schneider, "Generation of Highly Stable Millimeter waves with Low Phase Noise and Narrow Linewidth," IEEE Photon. Technol. Lett. 27(15), 1613–1616 (2015).

[26] S. Preussler, T. Schneider, "Attometer resolution spectral analysis based on polarization pulling assisted Brillouin scattering merged with heterodyne detection," Opt. Express 23(20), 26879–26887 (2015).

[27] S. Preussler, T. Schneider, "Stimulated brillouin scattering gain bandwidth reduction and applications in microwave photonics and optical signal processing," Opt. Eng. 55(3), 031110 (2016).

[28] E.P. da Silva, R. Borkowski, S. Preussler, F. Schwartau, S. Gaiarin, M.I. Olmedo, A. Vedadi, M. Piels, M. Galili, P. Guan, S. Popov, C.-S. Brès, T. Schneider, L. K. Oxenløwe, D. Zibar, "Combined Optical and Electrical Spectrum Shaping for High-Baud-Rate Nyquist-WDM Transceivers," IEEE Photonics Journal 8(1), 1–11 (2016).

Conferences

[1] T. Schneider, A. Wiatrek, R. Henker, S. Preußler, "Slow-Light ohne Pulsverbreiterung für optische Puffer und zur Dispersionskompensation," 10. ITG-Fachtagung Photonische Netze, Vol. 214, pp. 239–242, Leipzig, Germany, April 2009.

[2] A. Wiatrek, R. Henker, S. Preußler, and T. Schneider, "Comparative Investigation of Zero-Broadening Methods in Brillouin Based Slow-Light Systems," IET Irish Signals and Systems Conference 2009, Dublin, Ireland, Juni 2009.

[3] A. Wiatrek, R. Henker, S. Preußler, T. Schneider, "1.4 Bit Delay and Pulse Compression Based on Brillouin Optical Signal Processing," Slow and Fast Light, OSA Technical Digest (CD) (Optical Society of America, 2009), paper SMC4, Honolulu, Hawaii, USA, Juli 2009.

[4] A. Wiatrek, R. Henker, K. Jamshidi, S. Preussler, T. Schneider, "Numerische Berechnungsverfahren zur Simulation von Brillouin-basierten Slow-Light Systemen," Verhandlungen der DPG (VI) 45, 1/2010, Beitrag Q55.88, Hannover, Germany, März 2010.

[5] S. Preussler, K. Jamshidi, A. Wiatrek, R. Henker, T. Schneider, "Quasi-Lichtspeicherung mittels Zeit-Frequenz-Kohärenz," Verhandlungen der DPG (VI) 45, 1/2010, Beitrag

Q47.5, Hannover, Germany, März 2010.

[6] R. Henker, A. Wiatrek, <u>S. Preußler</u>, T. Schneider, "Langsames Licht auf der Basis der stimulierten Brillouin Streuung in photonischen Netzen mit unterschiedlich langen Standard-Einmodenfasern," 10. ITG-Fachtagung Photonische Netze, Vol. 222, pp. 183–186, Leipzig, Mai 2010.

[7] T. Schneider, K. Jamshidi, <u>S. Preußler</u>, "Quasi-Light-Storage eine Methode zur Speicherung optischer Datenpakete," 10. ITG-Fachtagung Photonische Netze, Vol. 222, pp. 265–270, Leipzig, Mai 2010.

[8] K. Jamshidi, <u>S. Preußler</u>, A. Wiatrek. R. Henker, J. Klinger, T. Schneider, "100 ns Quasi-Light-Storage of 8 Bit Data Sequences at 1 Gbps," Nonlinear Photonics, OSA Technical Digest (CD) (Optical Society of America, 2010), paper NME21, Karlsruhe, Germany, Juni 2010.

[9] <u>S. Preußler</u>, K. Jamshidi, A. Wiatrek, T. Schneider, "Almost Distortion Free Storage of 1Gbps/8bit Optical Packets for up to 100 Bit Lengths," ECOC 2010 - 36th European Conference and Exhibition on Optical Communication, Turin, Italy, September 2010.

[10] A. Wiatrek, <u>S. Preußler</u>, T. Schneider, "A Comparative Investigation of Broadening-Free Pulse Delay in Brillouin induced Slow-Light," 2nd Mediterranean Photonics Conference, Eilat, Israel, November 2010.

[11] T. Schneider, <u>S. Preussler</u>, A. Wiatrek, "Tunable storage of optical signals by Time-Frequency-Coherence," 2nd Mediterranean Photonics Conference, Eilat, Israel, November 2010.

[12] <u>S. Preußler</u>, K. Jamshidi, A. Wiatrek, T. Schneider, "Einfache, variable Speicherung optischer Daten bis zu 800ns," Verhandlungen der DPG (VI) 46, 1/2011, Beitrag Q8.2, Dresden, Germany, März 2011.

[13] K. Jamshidi, <u>S. Preußler</u>, A. Wiatrek, T. Schneider, "All optical Quasi Light Storage: achievments and limitations," 12. ITG-Fachtagung Photonische Netze, Vol. 228, pp. 200–203, Leipzig, Mai 2011.

[14] <u>S. Preußler</u>, K. Jamshidi, A. Wiatrek, T. Schneider; Einfache, variable, optische Datenspeicherung bis zu 800ns," 12. ITG-Fachtagung Photonische Netze, Vol. 228, pp. 208–211, Leipzig, Mai 2011.

[15] <u>S. Preussler</u>, K. Jamshidi, A. Wiatrek, T. Schneider, "Very Simple Tunable Optical Data Storage of 8Bit 1Gbps Data Packets Up to 500ns," Conference on Lasers and Eletrooptics/Quantum Electronicsand Laser Science and Photonic Applications, Systems and Technologies, paper CFP5, Baltimore, USA, Mai 2011.

[16] A. Wiatrek, K. Jamshidi, <u>S. Preussler</u>, T. Schneider, "Saturation and Delay in Broadband Brillouin Slow-Light," Slow and Fast Light, OSA Technical Digest (CD) (Optical Society of America, 2011), paper SLWA5, Toronto, Canada, Juni 2011.

[17] S. Preussler, K. Jamshidi, A. Wiatrek, T. Schneider, "Light Storage Enhancement by Reducing the Brillouin Bandwidth," Slow and Fast Light, OSA Technical Digest (CD) (Optical Society of America, 2011), paper SLWA4, Toronto, Canada, Juni 2011.

[18] T. Schneider, S. Preußler, K. Jamshidi, "Tunable Light-Storage for almost 1 Microsecond," Slow and Fast Light, OSA Technical Digest (CD) (Optical Society of America, 2011), paper SLMB5, Toronto, Canada, Juni 2011.

[19] A. Wiatrek, S. Preussler, K. Jamshidi, T. Schneider, "Managing the Resolution Bandwidth in Brillouin based Spectroscopy," 2011 International Students and Young Scientists Workshop Photonics and Microsystems, pp. 134-137, Cottbus, Germany, Juli 2011.

[20] S. Preussler, A. Wiatrek, K. Jamshidi, T. Schneider, "Methods for the Enhancement of the Storage Time in Quasi-Light-Storage," 2011 International Students and Young Scientists Workshop Photonics and Microsystems, pp. 103-105, Cottbus, Germany, Juli 2011.

[21] S. Preußler, A. Wiatrek, K. Jamshidi, T. Schneider, "Bandbreitenreduzierung der stimulilerten Brillouin Streuung für ultrahochauflösende Spektroskopie von optischen Signalen," Verhandlungen der DPG (VI) 47, 1/2012, Beitrag Q54.46, Stuttgart, Germany, März 2012.

[22] A. Wiatrek, S. Preussler, K. Jamshidi, T. Schneider, "Brillouin Gain Bandwidth Reduction down to 3 MHz in Standard Single Mode Fibers," Verhandlungen der DPG (VI) 47, 1/2012, Beitrag Q66.2, Stuttgart, Germany, März 2012.

[23] A. Wiatrek, S. Preussler, K. Jamshidi, T. Schneider, "Frequency Domain Aperture for Ultra-High Resolution Brillouin Based Spectroscopy," Conference on Lasers and Eletrooptics/Quantum Electronics and Laser Science and Photonic Applications, Systems and Technologies, JW4A.63, San Jose, USA, May 2012.

[24] T. Schneider, A. Wiatrek, S. Preußler, R.-P. Braun, M. Grigat, "Analysis of Ultra-High Bitrate Wireless Links as a Bridge for Optical Networks," 13. ITG-Fachtagung Photonische Netze, Leipzig, Mai 2012.

[25] K. Jamshidi, S. Preußler, A. Wiatrek, T. Schneider, "Wideband Electrically Tunable Dispersion Compensator/Producer," 13. ITG-Fachtagung Photonische Netze, Leipzig, Mai 2012.

[26] A. Mokhtari, S. Preußler, K. Jamshidi, T. Schneider, "Tunable Delay Line Based on Fourier-transformation and Linear Phase Modulation with High Time-Bandwidth Product," 13. ITG-Fachtagung Photonische Netze, Leipzig, Mai 2012.

[27] T. Schneider, A. Wiatrek, S. Preußler, R.-P. Braun, M. Grigat, "Maximum transmittable data rates for Millimeter-wave fixed wireless links," ELEKTRO 2012, pp. 94–98, Rajeck Teplice, Slovakia, May 2012.

[28] A. Mokhtari, K. Jamshidi, <u>S. Preussler</u>, A. Zadok, T. Schneider, "Highly Tunable Delay Line with Linear Phase Modulation and Optical Filtering," Advanced Photonics Congress, Silicon and Nano Photonics (IPR), Colorado, USA, June 2012.

[29] S.Treff, <u>S. Preussler</u>, T. Schneider, "Extension of an ordinary Network Analyzer as a cost-saving High Resolution Optical Spectrum Analyzer," 2012 International Students and Young Scientists Workshop Photonics and Microsystems, Wroclaw, Poland, July 2012

[30] <u>S. Preussler</u>, T. Schneider, A. Zadok, "Conquer the Limits of Brillouin Based Optical Spectrum Analysis," 2012 International Students and Young Scientists Workshop Photonics and Microsystems, Wroclaw, Poland, July 2012.

[31] <u>S. Preussler</u>, A. Wiatrek, K. Jamshidi, T. Schneider, "Increasing the Resolution of Optical Spectrometers for the Measurement of Advanced Optical Communication Signals," ECOC 2012 - 38th European Conference and Exhibition on Optical Communication, Amsterdam, Netherlands, September 2012.

[32] S. Treff, <u>S. Preussler</u>, T. Schneider, "Measuring the Spectra of Advanced Optical Signals with an Extension of an Electrical Network Analyzer," Optical Fiber Communication Conference (OFC), JW2A.20, Annaheim, USA, March 2013.

[33] <u>S. Preussler</u>, T. Schneider, "Bandbreitenreduzierung der stimulierten Brillouin Streuung in einem mehrstufigen, faserbasierten System," Verhandlungen der DPG, 1/2013, Beitrag Q29.4, Hannover, Germany, März 2013.

[34] <u>S. Preußler</u>, T. Schneider, "Increasing the storage time of Quasi-light storage by reducing the Brillouin gain bandwidth in a feedback system," 14. ITG-Fachtagung Photonische Netze, Leipzig, Mai 2013.

[35] <u>S. Preußler</u>, T. Schneider, "Proposal for the Tunable All Optical Storage of QAM Data Packets," Conference on Lasers and Eletrooptics/Quantum Electronicsand Laser Science and Photonic Applications, Systems and Technologies, JTu4A.96, San Jose, USA, June 2013.

[36] <u>S. Preussler</u>, T. Schneider, "Stimulated Brillouin Scattering Gain Bandwidth Reduction in Optical Fibers," 2013 International Students and Young Scientists Workshop Photonics and Microsystems, St. Marienthal, Germany, July 2013.

[37] <u>S. Preussler</u>, N. Wenzel, A. Zadok, T. Schneider, "Tunable generation of ultra-narrow linewidth millimeter and THz-waves and their modulation at 40 Gbd," 2013 IEEE International Topical Meeting on Microwave Photonics (MWP), Alexandria, VA, USA, Oct. 2013.

[38] <u>S. Preußler</u>, T. Schneider, "Bandbreitenreduzierung der stimulierten Brillouin Streuung in Monomodefasern," DPG Frühjahrstagung SAMOP, Beitrag Q30.18, Berlin, Germany, März 2014.

[39] S. Preußler, R.-P. Braun, M. Grigat, T. Schneider, "Erzeugung hochqualitativer Träger-wellen für zukünftige hochbitratige THz-Funkstrecken," 15. ITG-Fachtagung Photonische Netze, Leipzig, Mai 2014.

[40] T. Schneider, N. Wenzel, S. Preußler, "Generation of almost-ideally, sinc-shaped Nyquist pulse sequences with arbitrary bandwidth and repetition rate," 15. ITG-Fachtagung Photonische Netze, Leipzig, Mai 2014.

[41] S. Preußler, T. Schneider, "Tunable generation of high quality mm- and THz-waves for wireless communications with carrier frequencies up to several THz," World Telecommunications Congress, Berlin, June 2014.

[42] S. Preussler, N. Wenzel, A. Zadok, T. Schneider, "Generation of Flat, Rectangular Frequency Combs with Arbitrary Bandwidth and Frequency Spacing," Conference on Lasers and Eletrooptics/Quantum Electronicsand Laser Science and Photonic Applications, Systems and Technologies; STu3I.2, San Jose, Juni 2014.

[43] H. Al-Taiy, N. Wenzel, S. Preußler, J. Klinger, T. Schneider, "Ultra-Narrow Line-width, Stable and Widely Tuneable Laser Source for Coherent Optical Communication Systems," 40th European Conference and Exhibition on Optical Communication, Cannes, France, September 2014.

[44] S. Preußler, T. Schneider, "Erzeugung von flachen, rechteckförmigen Frequenzkämmen mit beliebiger Bandbreite und variablem Frequenzabstand," DPG Frühjahrstagung SAMOP, Beitrag Q 62.107, Berlin, Germany, März 2015.

[45] H. Al-Taiy, S. Preußler, T. Schneider, "Extra-narrow linewidth, stable and widely tunable extracted single laser line," DPG Frühjahrstagung SAMOP, Beitrag Q 40.5, Berlin, Germany, März 2015.

[46] H. Al-Taiy , S. Preußler, T. Schneider, "Methods for the Frequency Stabilization of High Quality Millimeter Waves with Low Phase Noise," 6th IEEE Student Conference, Hamburg, Mai 2015.

[47] S. Preußler, H. Al-Taiy, T. Schneider, "Generation and Stabilization of THz-waves with Extraordinary Low Line Width and Phase Noise," Conference on Lasers and Eletrooptics/Quantum Electronicsand Laser Science and Photonic Applications, Systems and Technologies; STu4H.6, San Jose, May 2015.

[48] S. Preussler, H. Al-Taiy, T. Schneider, "Optical Spectrum Analysis with kHz Resolution Based on Polarization Pulling and Local Oscillator Assisted Brillouin Scattering," 41th European Conference and Exhibition on Optical Communication, Valencia, Spain, September 2015.

[49] H. Al-Taiy, S. Preußler, S. Brückner, J. Schoebel, T. Schneider, "Generation of Highly Stable Millimeter waves with Low Phase Noise and Narrow Linewidth," IEEE Photonics Society Conference, pp. 98-101, Reston, Virginia USA, 2015.

[50] <u>S. Preussler</u>, A. Zadok, Y. Stern, T. Schneider, "Microwave Photonic Filters," German Microwave Conference 2016, Bochum, Germany.

[51] T. Schneider, <u>Stefan Preussler</u>, "Ultra-high resolution spectroscopy of optical frequency combs," SPIE Photonics West 2016, San Francisco, USA.

Patents

[1] T. Schneider, J. Klinger, <u>S. Preussler</u>, A. Wiatrek, S. Treff, "Netzwerkanalysator als Teil eines optischen Spektroskops," publication no.: EP 2664903 A3

[2] <u>S. Preussler</u>, T. Schneider, A. Wiatrek, R.-P. Braun, M. Grigat, "Kompensation des Dopplereffektes bei der Datenübertragung im Millimeter- und Teraherzbereich," publication no.: DE102012106958 A1

[3] R.-P. Braun, T. Schneider, <u>S. Preussler</u>, M. Grigat, "Übertragungsanordnung zum übertragen von Daten mit einer Trägerwelle im Terahertzbereich," publication no.: EP 2876824 A1

[4] <u>S. Preussler</u>, M. Grigat, T. Schneider, R.-P. Braun, "Datennetz in einer Passagierkabine zur Anbindung von elektronischen Endgeräten," publication no.: EP 2827513 A3

[5] R.-P. Braun, T. Schneider, <u>S. Preussler</u>, M. Grigat, "Verfahren zum Betreiben eines drahtlosen Datennetzes umfassend eine terahertzfähige Sendestation und ein terahertzfähiges, mobiles elektronisches Endgerät," publication no.: DE 102013101987 A1